Kids Are Consumers, Too!

Real-World Mathematics for Today's Classroom

Jan Fair

Mary Melvin

 Addison-Wesley Publishing Company

Menlo Park, California • Reading, Massachusetts

Wokingham, Berkshire, U.K. • Amsterdam • Don Mills, Ontario • Sydney

Acknowledgments

Special thanks to Palo Verde School, Tom Steege, Principal (Palo Alto, CA), where the photographs in this book were taken.

Teachers: Linda Ansell, Olive Borgsteadt, Theo Chang, Louise Waxham, Yvonne Westre

Children: Tom Addington, Scott Anderson, Philip Artates, Jon-Paul Barkhurst, Julie Berry, Myisia Brooks, Christopher Chan, Selby Cooper, Allison Danna, Lisa de Larios, Candy Epstein, Steven Farr, Susan Flynn, Ron Galant, Sheila Hsu, Timy Griffin, Khalid Hanafy, Twylah Harper, Becky Hayward, Samantha Jang, Alona Jasik, Jeff Kirby, Debbie Kass, Bryan Koski, Claudia Leiva, Shahrazad Livingston, Victor Mabutas, Anna Moore, Jennifer Morgan, Brent Nevius, Erik Nierenberg, Tom Parker, Alma Power, Christa Reese, Ron Rubin, Tina Ryge, Melissa Selbert, Shaka Tilghman

This book is published by the Addison-Wesley Innovative Division.

Illustrations: Jim McGuinness

All photographs provided expressly for the publisher by Wayland Lee, Addison-Wesley Publishing Company, Inc.

ISBN-0-201-20288-3

FG-AL-89

About the Authors

Jan Fair has used real-world experiences to teach mathematics in six states, in rural, suburban, and urban schools. She has authored numerous books and educational materials and is a speaker and consultant for school districts as well as community and parent groups throughout the United States and Canada. Ms. Fair is currently a math instructor at Allan Hancock College in Santa Maria, California, and is the director of AEM, Association for Education and Motivation, which conducts motivational and educational seminars nation-wide.

Mary Melvin has taught elementary students for more than twenty years—most recently in the McGuffey Elementary Laboratory School at Miami University, Ohio. Her particular interest in real-world experiences for learning mathematics has been evident in her elementary math classes and in her presentations at state and regional math meetings. Dr. Melvin currently teaches in the department of teacher education, Miami University, and is co-director of COMPUTERS IN THE CLASSROOM, an organization dedicated to helping teachers learn about educational applications of computers.

CONTENTS

MATH SKILLS COVERED IN THE BOOK

Charts and Tables (See also Graphing)
    Classifying
    Collecting data
    Reading and interpreting
    Recording
    Sampling
    Sorting

Decimals (See also Numbers and Numeration)
    Addition
    Subtraction
    Multiplication
    Division
    Averaging

Estimation

Fractions (See also Numbers and Numeration)
    Addition
    Subtraction
    Multiplication
    Division

Geometry
    Coordinate geometry
    Proportional drawing
    Shapes
    Spatial relations
    Symmetry
    Vocabulary

Graphing (See also Charts and Tables)
    Bar graphs
    Circle graphs
    Coordinates
    Line graphs
    Pictographs
    Scattergrams

Measurement (Metric and Customary)
    Area
    Estimation
    Linear
    Mass
    Perimeter
    Scale measurement
    Temperature
    Vocabulary

CURRICULAR AREAS RELATED TO ACTIVITIES

CURRICULAR AREAS

CHAPTER/ACTIVITY

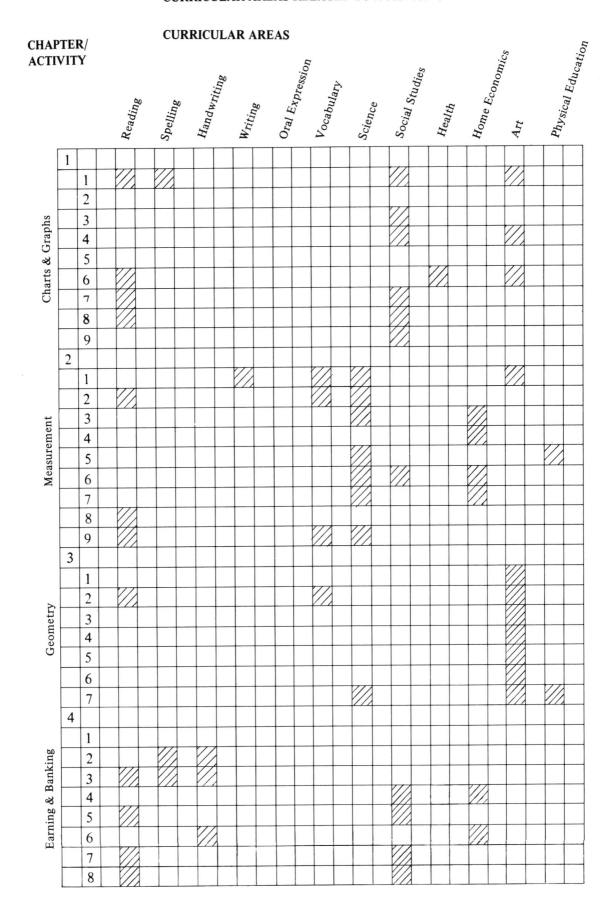

CURRICULAR AREAS RELATED TO ACTIVITIES

CURRICULAR AREAS

CHAPTER/ ACTIVITY	Reading	Spelling	Handwriting	Writing	Oral Expression	Vocabulary	Science	Social Studies	Health	Home Economics	Art	Physical Education
5 Shopping												
1												
2	X											
3	X	X	X									
4										X		
5										X		
6					X		X					
7	X		X									
8	X	X	X									
9	X									X		
10		X									X	
6 Eating												
1	X	X	X			X					X	
2	X											
3							X					
4							X			X		
5										X		
6										X		
7	X									X		
7 Traveling												
1	X							X			X	
2	X							X			X	
3								X			X	
4	X											
5	X											
6								X				
7	X							X				
8 Projects												
1	X			X		X	X			X		
2	X							X		X	X	
3	X			X				X				
4	X						X			X		
5	X			X				X		X		
6	X	X	X			X		X		X	X	

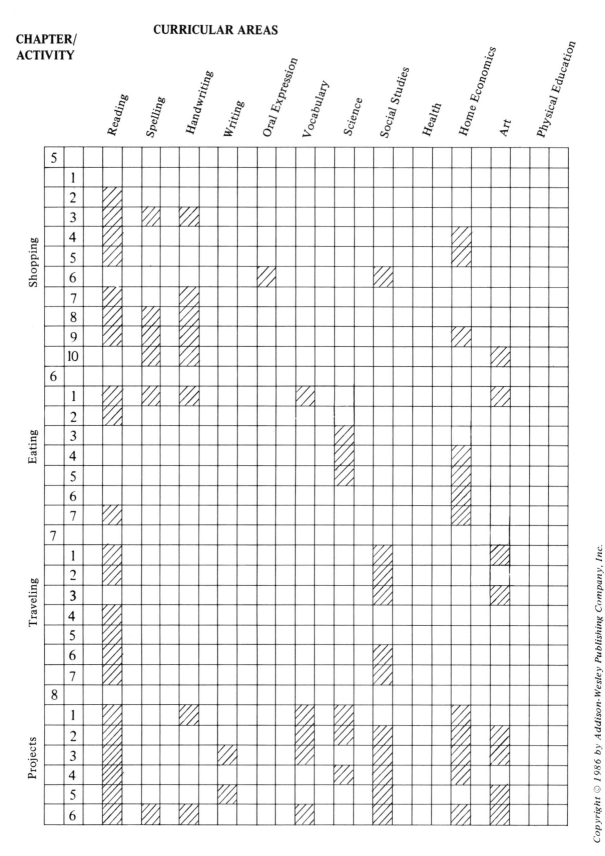

Introduction

No doubt about it—kids are consumers! They spend, save, and earn money. They buy snacks with their allowance, save their babysitting money for computer games, and deliver newspapers to earn enough for a new bike. They deal with measurement, estimation, and comparison. On television they're greeted by the newest toy, the biggest hamburger, the tastiest toothpaste, and the best deal on a bicycle. Our children deserve more than consumer education by TV. They deserve the chance to become wise consumers, to learn the skills needed to solve problems and to make decisions in the real world.

The heart of real-world mathematics, of being a careful consumer, is problem solving. And there's a growing emphasis on the importance of teaching problem-solving skills throughout the curriculum. The general public is making it very clear that they expect schools to teach the survival skills young people need to function successfully in the real world. And our professional organizations are emphasizing these same issues. For example, the National Council of Teachers of Mathematics recommends that problem solving be the focus of school mathematics.

This book is our response to these concerns. We recognize that the best way to help kids learn problem-solving skills is to get them involved in a realistic way. We also realize that many teachers are already putting a lot of effort into doing just that. The purpose of this book is to provide you with a rich collection of ideas and materials that will help you do an even better job of making your math curriculum real—and, as a result, help your students become wise consumers in the real world of their daily lives.

What You'll Find in This Book

This book is a collection of real-world math activities. They're not just simple math problems; they're the kind of real-world situations that we meet every day. Let's look at how the book is put together.

The Book's Organization

We've organized the math activities into two main categories. In Chapters 1, 2, and 3, the focus is on skill areas: Charts and Graphs, Measurement, and Geometry. The other activity chapters (4 through 7) focus on real-world math applications: Earning and Banking, Shopping, Eating, and Traveling.

The last three chapters have quite a different flavor. Chapter 8, Projects, describes six comprehensive projects designed to give your students in-depth opportunities to apply their real-world math skills! Chapter 9, Tips for Teachers, helps you plan the activities to suit your particular needs. Chapter 10, Materials and Resources, consists of suggestions and ideas for using the real-world materials (menus, maps, and so on) that you'll find throughout the book.

Organization of the Activity Chapters

Each of the activity chapters, 1 through 7, includes three parts. The first part is a page of introductory suggestions. On these pages, we've shared some of our special ideas about using the chapters.
The second part of each chapter is the collection of activities. Their format is described in detail below.

At the end of each chapter, you'll find a collection of Quick and Easy activities. These are five-minute fillers, the kind of activities that are perfect for those moments when it's too late to start a new lesson but too early to go to lunch!

Format of the Activities

All the activities are written in a simple, easy-to-use format. Each one starts on a new page and includes the following information:

GRADE LEVELS: The grade levels for which the activity is most appropriate (circled in the box at the top of the page).

TITLE AND DESCRIPTION: A phrase that tells the main idea of the activity, followed by a brief description of what happens in the activity.

Math Skills: The math skills that students use in the activity.

Curricular Areas: Other areas of the curriculum that are included in the content and/or process of the activity.

Materials: All items that will be used in carrying out the activity.

A personal note to you with some of our own thoughts about the activity.

PREPARATION: Description of things to be done before introducing the activity to the class.

DISCUSSION: General questions for the class discussion preceding the activity, to highlight the main ideas and concepts of the activity.

DIRECTIONS: Step-by-step instructions describing how to direct the activity in the classroom. We've written these in the actual words you might want to use. Instructions meant only for you are in parentheses.

VARIATIONS: Ideas for extending or using a different approach to the activity.

Underlying Themes

The main themes of this book are real-world problems and math applications. Because these ideas are so closely related to the lives of real people, we've also emphasized two underlying themes that relate to the human aspect of math class.

One of these themes is respect for the uniqueness of each child. Some activities are designed to encourage children to express their opinions, to name a favorite thing, or to describe something unique about themselves. In other activities, there are suggestions to help you create a warm, supportive environment for your students. And at the end of the book, you'll find award certificates for those occasions when you want to recognize something special about your students.

Our second underlying theme is promoting good relations among the school, the home, and the community. Because you're teaching about the real world, there'll be many times when you'll want your students to gather data at home, or to have parents help collect materials. For some lessons, you may want to invite guest speakers from the community. We hope you'll take advantage of these consumer-related opportunities for your students to interact with the adults in your community. The adults will love the chance to know more about what's going on at school; you'll end up with a strongly supportive group of people who'll be eager to help in any way they can; and your students will know that they live in a very special world!

Part 1.
Skill Area Activities

Charts and Graphs

Line Up! Activity 1-10c

1. *Restaurant Bar Graph*

2. *Bar Graph Money Game*

3. *What's All This Junk?*

4. *The Color of Your Clothes: A Scattergram*

5. *Average Daily Attendance*

6. *Pictographs for Breakfast*

7. *Charting the Average Price*

8. *Circle Graphs Make Prime-Time Viewing*

9. *Don't Get Caught in a Graphic Jam*

10. *Quick and Easy*
    A. *A Quick Birthday Graph*
    B. *What's for Lunch?*
    C. *Line Up!*
    D. *Menu Planning Scattergram*
    E. *My, How Times Have Changed!*

Charts and graphs are not only fun to do, but they'll make your room look great for open house! Besides that, your students will learn a wonderful variety of very important basic skills.

In the real world, we are confronted with massive amounts of information every day. Students need to be able to cope with the data they face as they read a magazine, flip through a catalog, shop in a store, or hear statistics on the six o'clock news.

The activities in this chapter suggest easy ways students can collect data—and even easier ways to organize, present, and interpret the information using charts and graphs. Your students will chart grocery prices, take a closer look at the treasures they've stuffed into their desks, and build a bar graph about their favorite snack foods. Such activities not only give them a chance to get personally involved, but also go a long way toward helping them to deal successfully with their world.

In this chapter, you'll find a variety of graphs. The easiest to do are the bar graphs and scattergrams. Circle graphs and pictographs are more challenging—but once your students get involved with them, they'll discover that graphs can be fascinating tools for recording and sharing information.

As you look over the activities in this chapter, choose those that suit your needs, skip those that don't. And while you're making choices, remember that you don't have to do these activities exactly as they're written. *You* know your students and *you* know how to create the best learning environment for them. The ideas in this chapter, along with your creative efforts, will be a super combination for introducing your students to the fun and challenge of charts and graphs.

Restaurant Bar Graph

Students make bar graphs to show their favorite local restaurants.

Math Skills:	making bar graphs
Curricular Areas:	reading • spelling • social studies • art
Materials:	1-cm graph paper (page 284) • crayons or markers

Kids have strong preferences about where they like to go to eat. This activity will certainly spark lively discussions of the pros and cons of the local restaurants.

DIRECTIONS:

1. We're going to make a bar graph to show our favorite places to eat in town. Let's list six of them on the board.

2. Please look at this list and decide which one of these six you like best. As I read the names, raise your hand to show your choice. (As you name them, record the number for each one.)

3. Make six columns on your paper—each one three cm wide. Write the name of one of the places to eat at the bottom of each column.

4. Moving upward along the left side of your paper, number the lines from 1 to _____ (one more than the largest number for any single restaurant on the board).

5. How many people chose the first place to eat on the board? On your graph, color in that number of spaces in the column for that place. Do the same for the other five places to eat.

6. Let's look at the results. Why is one place to eat more popular than another? What do you think the least popular place to eat should be doing to increase business?

VARIATION:

Make graphs showing student choices of just one kind of restaurant (for example, pizza places, fast food restaurants, coffee shops).

Bar Graph Money Game

In groups of four or five students, players take turns spinning an amount of money. They record this amount on a bar graph and race to the finish line.

Math Skills: making bar graphs • counting by fives • adding money

Materials: 1-cm graph paper (page 284) • game spinner (page 280) • paper clip and pencil

---

If you're looking for a good Friday-afternoon game, this is it!

PREPARATION: Each group of players will need one game spinner and one graph.

To make spinner:

1. Duplicate game spinner (page 280).

2. Fill in the five sections with one of the following sets of information:
   - 5¢; 10¢; 15¢; 20¢; 25¢
   - 5 pennies; 2 nickels; 1 nickel and 1 dime; 2 dimes; 1 quarter.

   (You could also paste pictures of the above denominations in each section.)

To make the graph:

1. Moving upward on the left-hand side of the graph paper, mark off each line in five-cent intervals. (5¢, 10¢, 15¢ . . . $2.00)

2. Draw a dark line along the top at $2. This is the finish line.

3. Divide the graph paper into equal vertical column sections. Each group will need as many sections as there are players.

4. Have each player put his name at the bottom of one of the columns.

DISCUSSION: Everybody takes a turn spinning the money game spinner. Who spun the least amount of money? You'll go first when we start playing.

DIRECTIONS: 1. First person spins and records the total amount shown on the spinner by shading the appropriate number of boxes in his column.

2. Play continues to the left.

3. Each time a player spins, he adds the amount he spun to his previous total by shading in the appropriate number of boxes.

4. The first person to reach $2 wins.

5. (Optional: Winner is the person who gets the highest amount in a given time frame, such as 30 minutes.)

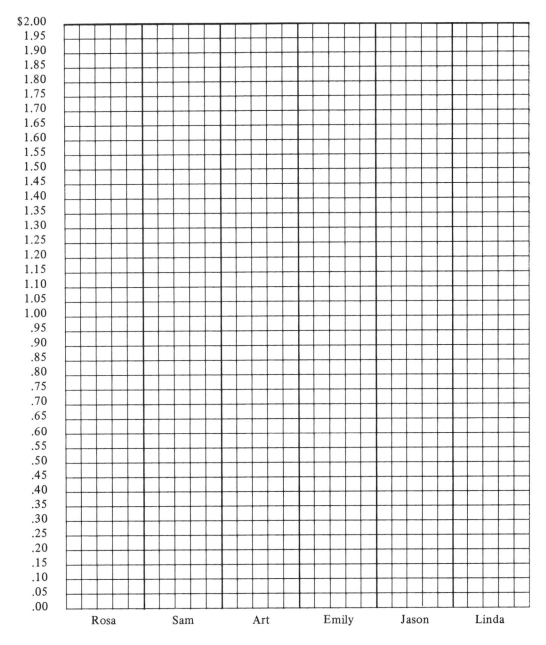

| | Rosa | Sam | Art | Emily | Jason | Linda |

What's all this Junk?

Students choose two to four items from their desks (or wherever they store their school supplies and materials), then decide on different ways to classify the items and record the information.

Math Skills: sorting • classifying • recording

Curricular Areas: social studies

Materials: items in students' desks, lockers, or other storage areas

---

This activity requires a willingness to accept the unexpected. After it's over, it may even serve as a motivation for getting those desks cleaned out.

PREPARATION: Be sure there's an area in the room where the students can sit in a circle (on the floor or on chairs).

DISCUSSION: What kinds of things do you keep as part of your school supplies? Do these items fit into any special categories?

DIRECTIONS:
1. Please take two things out of your desk (locker, tote tray, or whatever) and place them in this area so that everyone can see all of them.

2. If I asked you to sort this entire collection into groups, how would you sort them?
   (Encourage students to suggest different ways to sort the items.)

3. Which way do you think is the most logical way to sort them? Why?
   (Help the class decide on one way to sort the items: hard and soft, necessary at school or just for fun, by color, shape, and so on.)

4. If we needed to keep a record about the kinds of things you keep at school, how could we do it?
   (Using students' suggestions, help them agree on a suitable way to record this information: chart, bar graph, list of items, and so on.)

5. Would someone please draw an outline of the chart (or graph, table, and so on) on the board? Would someone else please record the information as we name our items one at a time?

6. What does this information tell us about the kinds of things you keep in your desks?

The Color of Your Clothes: A Scattergram

Students record data about the color of their clothing in a scattergram.

Math Skills:	sampling • making scattergrams
Curricular Areas:	social studies • art
Materials:	large newsprint paper or chalkboard

Guaranteed to intrigue your fashion-conscious students . . . and it's really easy to do.

PREPARATION: Draw a 4 x 4 box grid on newsprint or on the board.

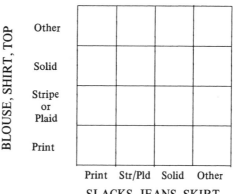

DISCUSSION: How can two sets of data be recorded in a single graph? When two events occur together, does that mean there's a correlation?

DIRECTIONS:

1. Look at the color of your clothing. Decide which of these words–print, stripes/plaid, solid color, or other–describes your slacks, jeans, or skirt. Which describes your shirt, blouse, or the top of your dress?

2. The students at Table A, please go to the board and put a small *X* in the box that describes both items of your clothing. (Continue until all students have been to the board.)

3. Is there any pattern on the scattergram?

4. Do the results imply any correlation between the colors of these two pieces of clothing? Can you explain these results? How might this information be used in the fashion industry?

VARIATION: Have students observe clothing in other classrooms or places. They can make scattergrams and share the information with the class.

Average Daily Attendance

Using actual (or fictional) data, students calculate the average daily attendance for several classes for the past week.

Math Skills: addition • division • averaging • making charts and tables

Materials: the daily attendance count for several classes for the past week
• large newsprint paper or copies of a small chart for each student (see *PREPARATION*).

This activity is easy to grade! Everyone uses the same data and solves the same problems. And, happily, it will help your students get a better understanding of why you're so careful about your attendance records.

PREPARATION: Prepare a large chart showing last week's attendance for several rooms. If you prefer, make a small chart and duplicate it so that each child will have a copy, or make a transparency for the overhead projector. (Perhaps an aide, a volunteer, or a student assistant could make the chart.)

SAMPLE CHART – AVERAGE DAILY ATTENDANCE

Week of November 10

		M	T	W	TH	F
	Rm 10:	25	18	17	22	24
FOURTH	Rm 11:	17	21	21	19	22
GRADE	Rm 12:	22	19	20	23	25
	Rm 13:	25	24	24	23	18
AVERAGE						
	Rm 101:	28	28	30	26	27
FIFTH	Rm 102:	29	29	29	28	29
GRADE	Rm 103:	24	29	30	30	28
	Rm 104:	31	33	30	34	29
AVERAGE						

DISCUSSION: Do you know how many students there are in the fifth grade (or use your own grade) at our school? Are all those students here every day? How many would you estimate are here each day? (Continue the discussion about attendance in another grade.) Why do you think teachers have to turn in their attendance every day? (Help students understand the relationship between attendance and school funds.)

Copyright © 1986 by Addison-Wesley Publishing Company, Inc.

DIRECTIONS:

1. Using attendance information from the school office, we can figure out the average attendance for each day.

2. On the board (on your paper) are the attendance figures for some classes for last week. How do we determine the average daily attendance for one grade level? (Add the attendance in all fourth-grade rooms for Monday; divide by the number of rooms.)

3. Each of you please calculate the average daily attendance for the grade levels shown on the chart.

VARIATION:

Have students make up a computer program to calculate average daily attendance for each grade level and for the whole school.

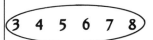

Pictographs for Breakfast

The class creates a pictograph illustrating how many pieces of toast they ate for breakfast the past three days. Then each student makes a pictograph showing how many pieces of fruit the class ate on the same days.

Math Skills:	making pictographs
Curricular Areas:	reading • health • art
Materials:	crayons or markers • paper

Students will enjoy designing their own symbols for recording data on a pictograph.

PREPARATION: List the names of the last three days in a column on the board. Think of a picture symbol you could use for toast.

DISCUSSION: What are some things you usually eat for breakfast?

DIRECTIONS:

1. One way to record information about the foods we eat is in a pictograph. Let's do one to show how much toast we've eaten in the last three days.

2. Let's count how much toast you ate for breakfast this morning—one hand for each piece of toast—and please stand if you ate more than two! (Have a student count and record today's toast on the board. Ask for the same information for yesterday and the day before.)

3. What picture symbol shall we use for the toast pictograph? (piece of toast, toaster) How many pieces of toast should each picture symbol represent? (2, 5, 10)

4. (Using the data on the board, have students determine how many picture symbols are needed to represent the toast eaten on each of the three days. For example: One picture of toast represents 10 pieces of toast; 35 pieces were eaten today. The pictograph needs 3½ pieces of toast.)

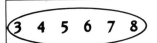

WEDNESDAY

TUESDAY

MONDAY

Key: = 10 slices of toast

5. Now that you've become pictograph experts, we'll collect more data and each of you can make your own pictograph. For these graphs, let's count how many pieces of fruit you've eaten in the same three days. (Collect data as for toast.)

6. Using this information, please design your own pictograph. (Encourage students to choose an interesting picture symbol.)

Charting the Average Price

Using information from grocery ads, students calculate the average price for four staple items on a given day; they do the same thing a month later and compare the average prices.

Math Skills: addition • division • averaging • making charts and tables

Curricular Areas: reading • social studies

Materials: large newsprint paper • grocery ads from three or more stores (for the same day or week)

Once the students get started on this, they may become regular fans of the weekly grocery ads—and they'll surely have a better idea of where the family food money goes.

PREPARATION:
1. Identify four staple grocery items that are regularly advertised in your area (for example, milk, bread, flour, eggs).

2. Display the grocery ads from at least three stores in the area. (You may want to write each store's prices for each of the four staples on the board.)

3. Prepare a display chart.

		ITEMS			
		Milk	Bread	Flour	Eggs
S T O R E S	1. MIKE'S MARKET				
	2. PAY-NO-MORE				
	3. TINA'S 10-2				
	AVERAGE PRICE				

4. If your class hasn't done any averaging lately, you might want to review the process using items from the grocery ads.

DISCUSSION: How similar are the food prices at different supermarkets? Do the prices change in more or less the same way in all stores?

DIRECTIONS:

1. We're going to take a careful look at some prices in our local supermarkets. On your paper, please draw a chart like the one up here.

2. Use the information in the ads (or on the board) to find this week's prices for the items on your chart, then fill them in.

3. For each item, calculate the average price for the three stores and record it on your chart.

4. Does one store usually have the lowest or highest prices? Which price is farthest from the average? How much more (or less) is it?

5. (Students' charts should be saved to be used in the continuation of this activity one month later. At that time, new grocery ads are posted in the room, and students follow steps 1 through 3 again. Students then compare the new averages with the old ones and suggest reasons for any differences in averages. This activity can be continued on a monthly basis for an extended period of time.)

VARIATION:

Students can make up a computer program to calculate the average prices. New information can be added each month to keep a long-term record of average prices of grocery items.

Circle Graphs Make Prime-Time Viewing

Students make a circle graph to show what kinds of TV programs are broadcast in a given four-hour period.

Math Skills: keeping track of time use • making circle graphs

Curricular Areas: reading • social studies

Materials: weekly TV schedule for each student (or copies of the schedule for one four-hour time period)

Once students learn the basics of making circle graphs, they may enjoy making them to show how they use their time on different days and in different settings. For some, it could be a real eye-opener about why certain things do—or don't—get done as intended!

PREPARATION: On chart paper, draw a blank circle graph. Divide the circle into 16 equal segments. One way to do this is:

1. Fold the paper horizontally, fold the paper vertically, then unfold it.

2. Use the point where the lines cross as the center of the circle. Draw a circle, using a compass or a string and pencil.

3. On the circle, draw marks to divide the circle into fourths. Estimate and draw marks to divide into eighths, then sixteenths. (These same directions will help guide students when they draw their own circle graphs.)

DISCUSSION: What kinds of TV programs are generally shown during weekday prime time—6 to 10 p.m.? (News, game shows, sit-coms, drama, music.) What kinds of programs are shown during other four-hour periods, such as 8 to 12 Saturday morning? Late Friday night from 10 to 2? Tuesday from 10 a.m. to 2 p.m.? What kinds of programs get the most time during prime time? On Saturday morning? Late Friday night? Midday during the week?

DIRECTIONS:
1. A circle graph is one way to show how a given period of time is divided. Let's make a circle graph to show how one of our TV stations schedules the four-hour time period from 6 to 10 on Tuesday.

2. I've marked this graph to show 16 equal parts. When we use it to show a four-hour block of time, how much time will each of these segments represent? (fifteen minutes)

Copyright © 1986 by Addison-Wesley Publishing Company, Inc.

3. Please look at your TV schedules and find the schedule for Channel 2 (or other station) from 6 to 10 on Tuesday evening. What kinds of shows are there? (news, local events, drama, musical variety, and so on)

4. How much time is used for each of these kinds of programs? (List them on the board: News—1½ hours; Drama—1 hour; and so on.)

5. How can we show this information on the graph? (1½ hours for news—six segments; an hour for drama—four segments; and so on.) Will someone please color in and mark the time for news? for drama? (and so on.)

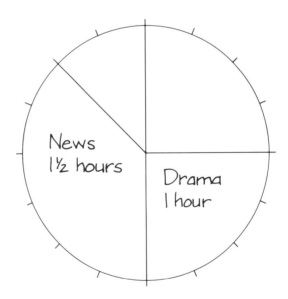

6. Using your TV schedules, please graph the kinds of programs on one of the other channels from 6 to 10 on Tuesday.

7. (For more variety, you might have students each choose a different four-hour time period or a different station for their own graphs.)

VARIATIONS:

1. Encourage students to graph the kinds of programs they watch at home when they have the TV on for a four-hour period of time.

2. Students can make circle graphs to show how they use their time for 24 hours. They may need help to figure out how to divide the circle into 24 equal segments.

Don't Get Caught in a Graphic Jam

Students list the methods of transportation they used to get to school today and then graph the information.

Math Skills: making graphs

Curricular Areas: social studies

Prerequisites: knowledge of at least one kind of graph: bar graph, line graph, circle graph, pictograph, scattergram.

Materials: 1-cm graph paper (page 284) • crayons or markers

Do some of your students walk to school and others ride in cars or buses? Then this is the activity for you. When it's finished, the graphs will look great on your bulletin board, in the principal's office . . . or even in the bus garage!

DISCUSSION: What kind of graphs have we learned to use? Why are there so many different kinds of graphs?

DIRECTIONS: 1. How did you get to school today? Did you ride your bike or skateboard? How many walked to school? Did anyone ride the subway or come on the school bus or in a car?

2. (List each method of transportation on the board: bike, bus, feet, and so on. Next to each, write the number of students in the class that used that method today.)

3. Think about the kinds of graphs we have studied. Choose one to use for this information. (Encourage the students to discuss and experiment with rough drafts before they try to do a final version of their graph.)

4. (Optional: If you've studied different types of graphs, you may want to divide the class into groups and assign one type of graph to each group. When the graphs are completed, have each group comment on the strengths and weaknesses of their graphs for showing this information.)

VARIATION: Students who have some knowledge of computer graphics might enjoy the challenge of writing a program to create their graph(s) on the computer.

Quick and Easy

A. A QUICK BIRTHDAY GRAPH

1. (List the twelve months on the board.)

2. In what month do you think the most people were born?

3. (Record the number of students whose birthdays occur in each month.)

4. How can we graph this information? (bar graph, pictograph, chart, and so on)

B. WHAT'S FOR LUNCH?

1. (List the items in today's school menu on the board.)

2. (Have students select—silently—one item that they would like to replace.)

3. (Beside the menu items, tally the times that item was chosen to be replaced.)

C. LINE UP!
(See photo, page 3.)

1. (Here's a different way to get your class to line up for an assembly, for lunch, or to go home! Ask students to name their favorite type of family vehicle—pickup truck, station wagon, van, compact, sedan, sports car, and so on. Write them across the board, listing each type only once.

2. We're going to make a human graph to show how many families have each kind of vehicle. Those who named a van please come stand in front of the word <u>van</u> on the board. Those who named pickup truck stand in front of that word. (And so on.)

3. Which kind of vehicle has the most people in front of it? Which has the fewest? Are there any lines that have the same number of people in them?

4. The compact line can go to lunch. Sports cars start your engines! You're next. . . .

D. MENU PLANNING SCATTERGRAM

1. (On the board, draw a scattergram showing four breakfast cereals across the bottom and three fruits up the side. You may want to have students name their favorite cereals and fruits.)

F	BANANAS				
R	RAISINS				
U	PEACHES				
I		CORN FLAKES	RICE FLAKES	OAT FLAKES	WHEAT FLAKES
T **S**					
		C E R E A L S			

2. Suppose we were going to advertise breakfast cereals served with fruit on them. Which combinations of fruit and cereal do you prefer? Each of you in the first row please come up and put an X in the box for your preference. (Have all students mark their preference.)

3. Which combination was chosen most often? Would that be the best combination to use in a cereal advertisement? Why or why not?

E. MY, HOW TIMES HAVE CHANGED!

1. (Ask students to name goods and services that are available today but weren't fifty years ago.)

2. (List the responses on a chart—one column for goods, one for services.)

3. (Reverse the question and ask for goods and services that were available fifty years ago, but aren't available now. List them on a second chart.)

4. (Encourage students to ask grandparents and other older persons for suggestions about additional goods and services to add to the second chart.)

Measurement

Grocery Grams Galore   Activity 2-6

1. *Measurement Unit of the Week*
2. *A Metric Schoolroom*
3. *A Watched Pot Always Boils*
4. *Rugs and Doormats*
5. *Pace, Place, and Calculate*
6. *Grocery Grams Galore*
7. *Kitchen Calculations*
8. *Weigh-Out Catalog Calculations*

9. *The Metric Marketplace Game*
10. *Quick and Easy*
    A. *Measure Your Guess*
    B. *What Shall I Wear?*
    C. *Metric Merchandise Measure*
    D. *Supermarket Sizes*
    E. *Carpet for Your Secret Hideaway*
    F. *Paint a Wall*
    G. *Helping Others Metrically*

21

There's no doubt that being a wise consumer involves understanding measurement. Here are some activities to help your students do just that. And don't worry about the metric activities. They're designed for the novice, so, if necessary, you can learn along with your students!

The best way to learn to deal successfully with measurement—is to measure! Take a look at the first two activities. They're full of easy-to-do ideas that will keep measurement alive in your classroom all year long. When you want to focus on just one unit of measurement, look through the other seven activities to find one that meets your needs. And, by the way, in addition to the activities in this chapter, you'll find many activities that involve measurement throughout this book.

We've kept the materials you'll need to a minimum. Measurement equipment should not be a problem, as most schools have some— somewhere! Finding it, however, may involve a little detective work. A good first try, of course, is asking other teachers in your building. Another good source may be the local high school science or math teachers.

As you inch (centimeter?) your way along in this chapter, we hope it will measure up to your expectations.

Measurement Unit of the Week

Every week a different measurement unit is highlighted. Students list items that measure approximately the weekly measurement unit.

Math Skills: measurement—linear, mass, area, volume • estimation • measurement vocabulary

Curricular Areas: writing • vocabulary • science • art

Materials: large newsprint paper • (optional: measuring equipment • rulers—page 281)

This easy activity is guaranteed to help you and your students become better estimators of measurement units—metric as well as customary.

DISCUSSION: Our weekly measurement unit is the meter. Can you think of an item that is approximately one meter wide? one meter tall? one meter long?

DIRECTIONS: (Monday)

1. On this chart, let's list consumer items that measure approximately this week's measurement unit. Every day we'll add to this list. (Examples: cm—paper clip width, diameter of Eric's fingernail; pound—dictionary, Stephanie's boot.)

(Tuesday)

2. Here are five items on the chalkboard ledge. Which measures more (less) than this week's measurement unit?

(Wednesday)

3. Find a picture in a magazine that depicts an item the size of our measurement unit.

(Thursday)

4. Play Guess the Item. One person thinks of an item the size of the measurement unit. She tells everyone the kind of place it would be found (for example, in the living room, on the playground, at the toy store). The game proceeds like 20 Questions.

(Friday)

5. Write a story or draw a picture of items that measure the same as the weekly measurement unit.

6. (Continue to carry out similar activities using a different unit of measurement each week.)

A Metric Schoolroom

Students label items in the room with metric measurements. The labeling can be introduced as a group activity, then continued informally over a period of time.

Math Skills: metric vocabulary • estimation

Curricular Areas: reading • vocabulary • science

Materials: index cards • crayons or markers • tape • metric measuring equipment (For centimeter rulers, see page 281.)

Once this activity is under way, you'll have metric labels all around the room. Visitors will enjoy seeing what your class is doing—and they just may learn a little more about metric measurement themselves!

PREPARATION:

1. Arrange the measuring equipment on three tables: one for linear, one for volume, and one for mass. Each table should also have approximately 20 cards, several crayons or markers, and some tape.

2. Provide a bulletin board area for displaying small items taped to cards. (Students can help decorate and label the bulletin board.)

DISCUSSION:

What are the basic units of metric measurement? (meter, liter, gram) How do we measure and describe quantities less than a meter or liter? (Divide by tens, hundreds; use prefixes *deci, centi*.) How do we describe more than the basic unit? (In multiples of ten; use prefixes *deka, hecto, kilo*.) In our room what objects would be measured by meters? by liters? by grams or kilograms?

DIRECTIONS:

1. Today we're going to label some objects in the room. The labels will tell the metric measurement of the objects. (With the class, measure some objects and label them.)

2. (Divide the class into thirds and assign each group to one of the measuring tables.)

3. Take a minute to look around the room and decide on the first object you'd like to measure. Use the equipment at your table to measure the object, then label it.

4. Please choose three or four objects to measure and label.

5. (If possible, keep the metric measuring equipment handy for several days. Students can make additional labels when they have free time.)

       Chapter 2: Measurement

6. Please look at the metric labels around the room. What is the longest object that's labeled? The shortest? What's the difference in length? What weighs closest to 1 kilogram? What would be twice as heavy? half as heavy? Which container holds half a liter? Is there a container that holds approximately 100 mL? Is that more or less than a liter?

A Watched Pot Always Boils

The class will measure the temperature of water as it changes from room temperature to boiling.

Math Skills: telling temperatures—Celsius or Fahrenheit • (optional: line graphs)

Curricular Areas: science • home economics

Materials: hot plate • pan • liter or quart measure • tap water • thermometer • demonstration thermometer showing both Celsius and Fahrenheit • stopwatch or clock with minute hand

---

This is a good lesson to use in preparation for cooking activities.

PREPARATION: Set up the hot plate in a safe place where it is visible to the class. Turn it on so it will be hot by the time the discussion is finished.

DISCUSSION: What is the temperature of our room? What's the temperature of boiling water? At what temperature does water freeze? (If you have both Celsius and Fahrenheit thermometers, you may want to encourage responses in both scales.)

DIRECTIONS:
1. Eric, will you please put about half a liter (or half a quart) of tap water in this pan?

2. What do you think the temperature of the water is when it comes from the faucet? Let's use the thermometer to check. Stephanie, will you please record the temperature of the tap water?

3. How hot will the water get when we put it on the hot plate? How long do you think it will take to reach its hottest temperature?

4. After the water has been on the hot plate for one minute, Jan, will you please check and record the temperature?

5. (Have students read and record the temperature on the board every minute. Include two or three readings after it reaches boiling so that students can observe that the temperature stabilizes.)

6. How long did it take for the water to reach 100 (or 212) degrees? Why doesn't the temperature go any higher? Did the temperature increase at a regular rate? (Illustrate with a line graph.)

7. (Repeat the activity using a lid on the pan. Then add salt to the water for the third round.)

VARIATION: Have students make up a computer program for a line graph to illustrate the information.

Rugs and Doormats

Students will calculate the area and perimeter of rugs in their homes. They will then discuss their results.

Math Skills:	calculating perimeter and area • measurement—linear
Curricular Areas:	home economics
Materials:	meter/yardsticks or tape measures • (optional: rulers—page 281)

This activity is full of surprises. Your students will discover that although the perimeter of two figures is the same, the area may not be.

PREPARATION: Have students measure the length and width of a rectangular rug, doormat, or bathroom rug at home. Indicate which units of measurement they are to use—metric or customary.

DISCUSSION: I sewed fringe around two of my bathroom rugs. For each rug, I used the same amount of fringe. Would you think these rugs each cover the same amount of space?

DIRECTIONS:

1. Stuart, please tell us the length and width of a rug you measured at home.

2. Let's all calculate the area of this rug. Let's also find the perimeter. (Practice finding the area and perimeter of several rugs.)

3. Now, calculate the area and perimeter of your own rug. (If some students didn't bring dimensions of a rug, write an example on the board for them to use.)

4. (Make three columns on the board. Label them Length and Width, Area, and Perimeter.) Let's list the length and width, the area, and the perimeter for the rugs of the people at this table.

5. Now that we have this information on the board, does anyone at another table have a rug with the same area as one of these? Are the length and width of yours the same?

6. Does anyone have a rug with the same perimeter as one of these? Are the length and width the same? Is the area the same?

7. What can we say about the relationship of the area to the perimeter of rectangles?

Pace, Place, and Calculate

Students will learn the length of their paces (in meters or yards), then use that length to measure distances.

Math Skills:	measurement–linear • counting • multiplication • division • estimation
Curricular Areas:	science • physical education
Materials:	Activity Sheet 2-5 • tape measure • comfortable shoes(!)

Estimating short distances will become much easier after you do this one. Enjoy the walk.

PREPARATION: Find a distance in or near your room that measures about 15, 20, or 30 meters or yards. (Down the hall, in the multipurpose room, on the playground.)

DISCUSSION: What is a pace? (The distance covered in 1 step.) In what ways would it be helpful to know how long your pace is?

DIRECTIONS:

1. Today, let's measure our paces. To do that, we'll go to _____ .

2. From this wall to that wall is _____ meters (or _____ yards). That's the distance we're going to pace.

3. Bob, Carol, Ted, and Alice, please pace that distance and count as you go. (Be sure they count "one" each time their foot hits the ground. Then send the other students to pace in groups of four or five.)

4. (Someone records the paces for each student. You may want to have students pace the distance three or four times to get a more accurate estimate of the length of their pace.)

5. Return to the classroom and calculate the length of your pace. (Example: distance divided by number of paces = 15 m ÷ 33 paces = .45 m or 45 cm.)

6. Now that you know your pace, you can find distances. For example, my pace is 45 cm and it's 40 paces to the teachers' lounge. How far is it? (1800 cm or 18 m.)

7. (Send the students off in small groups to pace off some distances —for example, from classroom door to: principal's office, bathroom, drinking fountain, cafeteria.)

8. (Students calculate these distances and compare their results.)

9. (Give students Activity Sheet 2-5 to complete.)

Pace, Place, and Calculate

Pace Length —————————————————————————————

Record the paces you take when you go places such as around the block, down a supermarket aisle, across a parking lot, to school, to the bus stop, or around a baseball diamond.

Place	Number of Paces	Distance
_____	_____	_____
_____	_____	_____
_____	_____	_____
_____	_____	_____
_____	_____	_____
_____	_____	_____
_____	_____	_____
_____	_____	_____
_____	_____	_____
_____	_____	_____
_____	_____	_____

Does the distance you have to walk determine what you do? Where you shop? Where you go?

Grocery Grams Galore

Students estimate the weight of common items from the supermarket. They then weigh the items and compare estimated and actual weights. (*See photo, page 21.*)

Math Skills:	metric measurement—mass • estimation
Curricular Areas:	social studies • science • home economics
Materials:	Activity Sheet 2-6 • balance scales and metric weights • collection of three to five items purchased at the supermarket (such as rice, flour, cereal, popcorn, crackers, bread, onions, potatoes, bananas, apples, zucchini, cabbage, nuts, cheese)

Estimating weight is an extremely useful consumer skill. Students enjoy the challenge of learning to estimate—for some, this may even become a favorite activity for indoor recess!

PREPARATION: This activity is easiest to manage as a learning station or small group activity. The groceries, balance scales, and weights can be set out on a table. Copies of Activity Sheet 2-6 can be placed in a folder on the table.

DISCUSSION: How does a gram compare to a pound? (much, much less) Even though the gram is the basic unit, which unit is often used to measure items such as butter, flour, and fruits? (kilogram) How much do you think one onion would weigh? this sack of flour? a package of nuts? a bag of ten apples?

DIRECTIONS: (Have students complete Activity Sheet 2-6.)

VARIATIONS:

1. Make a large chart to keep a record of items that weigh approximately 100 g, 500 g, and 1 kg. Label three columns on a large chart with these amounts. Ask students to look around their kitchens for other grocery items that weigh these amounts, and have them list these items on the chart. Encourage them to get their parents involved in estimating metric weights.

2. As the lists on the chart grow longer, bring some of the items to class so students can check their actual weights.

Grocery Grams Galore

MATERIALS NEEDED: Balance scales and metric weights
Collection of food items from the
supermarket

DIRECTIONS:

1. Examine the groceries. Find items that you estimate to weigh approximately 100 grams, 500 grams, and 1 kilogram. Write these items in the columns below.
2. Weigh each of the items and record their actual weights below the names of the items.
3. Look for your best estimates. Circle your best estimate in each column.

ESTIMATED WEIGHTS		
100 grams	500 grams (½ kilogram)	1 kilogram (1,000 grams)
lemon (112 g)	2 onions 440 g	cabbage 1,365 g

Kitchen Calculations

Students will calculate different ways to express the same measurements (for example, 1 cup = ½ pound). They will estimate measurements in relation to familiar quantities.

Math Skills: customary and metric measurement—volume • multiplication • division • estimation

Curricular Areas: science • home economics

Materials: Activity Sheet 2-7a • (optional: Activity Sheet 2-7b • quart and liter containers • pound and 500 mL or 500 g containers)

Knowing the relationships between different kitchen measurements is a very helpful skill for cooks of all ages!

DISCUSSION: Suppose you need milk for a recipe. You only have ¼, ½, and 1-cup measuring containers. How can you measure a pint? a quart? four tablespoons?

DIRECTIONS: (Have students complete Activity Sheet 2-7a.)

VARIATION: Give students Activity Sheet 2-7b. In this one, students learn the relationships between customary and metric measurements.

Kitchen Calculations Customary Equivalents

HELPFUL HINT: A pint's a pound the world around.

1 pint	= 2 cups (c)	1 pound (lb)	= 2 cups
1 cup	= 8 ounces (oz)	1 pound	= 16 ounces
1 ounce	= 2 tablespoons (tbsp)	1 pound	= 32 tablespoons
1 tablespoon	= 3 teaspoons (tsp)		

MARGARINE:

¼ lb =

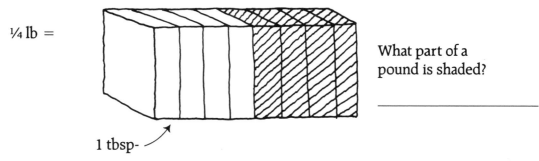

1 tbsp-

What part of a
pound is shaded?

_____

How many ounces in ¼ pound of margarine? _____

How many ounces in 1 tablespoon? _____

LIQUID

1 cup	= _____	ounces
½ cup	= _____	ounces
¼ cup	= _____	ounces
⅛ cup	= _____	ounces
2 cups	= _____	ounces
1 pint	= _____	cups
1 pint	= _____	pounds
1 pint	= _____	ounces
1 pint	= _____	tablespoons

1 cup	= _____	tbsp
½ cup	= _____	tbsp
¼ cup	= _____	tbsp
⅛ cup	= _____	tbsp
2 cups	= _____	tbsp
1 quart	= _____	pints
1 quart	= _____	cups
1 quart	= _____	pounds
1 quart	= _____	ounces
1 quart	= _____	tablespoons

Kitchen Calculations:

Metric and Customary Relationships

1 liter (L) is a little more than a quart.

1. 1 L = _____ mL

2. 500 mL is a little more than _____ .

3. 250 mL is a little more than _____ .

MARGARINE:

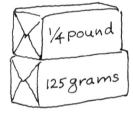

500 grams (g) is a little more than 1 pound.

4. 250 g is a little more than _____ pound.

5. 125 g is a little more than _____ pound.

500 milliliters (mL) is a little more than 1 pint.

6. 1000 mL is about the same as _____ pints.

7. 1000 mL is about the same as _____ quarts.

8. 500 mL is about the same as _____ cups.

9. _____ mL = 1 cup.

10. _____ mL = ½ cup.

THE MAGICIAN'S CORNER:

$$1 \text{ mL} = 1 \text{ g}$$

So... 125 mL = _____ g
       250 mL = _____ g
       500 mL = _____ g
      1000 mL = _____ g

MAGIC! 1000 mL = 1000 g = 1 kilogram!

Chapter 2: Measurement

Weigh-out Catalog Calculations

Students will calculate shipping weights for mail order purchases.

Math Skills: calculating with pounds and ounces • addition • subtraction • multiplication • division

Curricular Areas: reading

Materials: copies of catalog page (page 41) • order forms (page 42)

As long as most mail order items are measured in pounds and ounces, students will need slow, deliberate practice in calculating shipping weights for catalog orders.

PREPARATION: This activity works best when done over a period of several days. A sequence of three lessons is suggested here. You may want to review changing ounces to pounds before doing the activity.

DISCUSSION: When ordering by mail, why is it sometimes necessary to add an additional fee for shipping charges? How are shipping charges determined? (flat fee, weight, distance it must go, and so on)

DIRECTIONS: Lesson 1 (Adding pounds and ounces):

1. Please look at your catalog page. What can you find that weighs less than a pound? How much does it weigh? What items weigh between one and two pounds? more than five pounds? (and so on)

2. Please choose three items on your catalog page. On your order form, list them in the Name of Item column. Put the weight of each in the Shipping Weight column.

3. To find the total weight, add the pounds, then add the ounces. Now change the ounces to pounds and fill in the Total Weight in Pounds column.

Lesson 2 (Multiplying pounds and ounces):

1. (Name an item on the catalog page.) How much does it weigh? How much would two of them weigh? three of them? (Have students calculate weights for two or more other items.)

2. I'm looking at an item. Three of them would weigh _____ . What's my item? (Repeat several times; then have students make up similar problems.)

Lesson 3 (Multiplying and adding pounds and ounces):

1. Please cut out three shipping weights from your catalog page; cut out the pictures, too, if you want to.

2. Paste the weights and pictures at the top of a blank piece of paper. This time, you get to make up the problem!

3. Think of a problem about your three shipping weights. Try to make it include at least one chance to multiply and one to add. Write your problem on your page. Solve it and give your answer to the teacher.

4. (Students exchange papers and solve each other's problems.)

VARIATION: Students might like to work on computer programs to do one or more of the following:
  - Change ounces to pounds
  - Change pounds to ounces
  - Multiply a given weight by 2 or more
  - Find the total weight for an order

Catalog Page
(Use for Activities 2-8 and 5-8.)

BASKETBALL

BL108   1 lb 7 oz $14.95

TENNIS RACQUET (with cover)

SP 633   DELUXE nylon strung
              1 lb 2 oz $39.50

SP 643   REGULAR unstrung
              15 oz $19.75

SP 833   TENNIS BALLS (can of 3
                 extra duty)
              8 oz $4.25

SOCCER BALL

BL 41   black & orange
             1 lb 6 oz $13.95

BL42   black & white
             1 lb 4 oz $13.95

SWEATBANDS

One size fits all!

Colors: Red, White, or Blue

CL 909   Headband   2 oz $2.50

CL 910   Wristbands (set of 2)
                                 2 oz $2.50

CL 911   Set of head & wrist
                                 4 oz $4.00

SOCCER SHIN GUARDS

BLX 40   . . . . $8.00

Sizes: S (ages 6–9)   3 oz
             M (ages 10–14)   4 oz
             L (ages 14–adult)   5 oz

JUMPROPES

For exercise and fun!

Colors: White, Brown, Orange

SP 133   Small (15 oz) $2.50

SP 134   Medium (1 lb) $4.00

SP 135   Large (1 lb 1 oz)   . . . $7.50

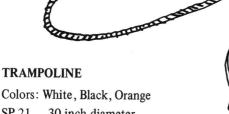

TRAMPOLINE

Colors: White, Black, Orange

SP 21   30 inch diameter
             17 lbs $47.50

SP 23   36 inch diameter
             21 lbs $66.75

Catalog Order Form
(Use for Activities 2-8 and 5-8 and Project 8-6.)

CATALOG ORDER FORM

DATE OF ORDER: _____

SHIP TO:

Name _____

Address _____

City _____

State _____ | Zip Code _____ | Office Use _____

BILL TO:

Name _____

Address _____

City _____

State _____ | Zip Code _____ | Office Use _____

Page	Item Number	How Many	Name of Item	Color	Size	Price Each	Total Price	Shpg. Wt.

MAIL ORDER TO:

	Total Lbs.	Total Oz.
Total for Goods		
Tax	Total Wt. in Pounds	
Shipping Costs		
Total		

The Metric Marketplace Game

Students move around a playing board by deciding whether a specific item is sold by the meter, gram, or liter. The game can also be played with the focus on just one type of measurement. (For example, players decide if items are sold by the milligram, gram, or kilogram.)

Math Skills: metric vocabulary • one or more of the following: linear, mass, volume, or area units of measurement

Curricular Areas: reading • vocabulary • science

Materials: game board (page 282) • playing piece for each student (coin, ring, bean) • index cards or slips of paper (see *PREPARATION* below for Purchase Cards) • (optional: old magazines or catalogs)

Here's a simple game that helps everyone learn about metric units used when buying things. It's easy to learn and fun to play.

PREPARATION: To make game board:

1. Duplicate game boards (page 282) for each group of two or three students. (Larger boards can be made by making a transparency and tracing around a projected image on tag board.)

2. Students fill in each space on the game board with one of these metric units of measurement: METER, GRAM, or LITER. There are 23 spaces that need to be filled in, including the circle and diamond spaces, the spaces that say IF YOUR MOVE FINISHES HERE, MOVE ACROSS, and the last space, which says FINISH.

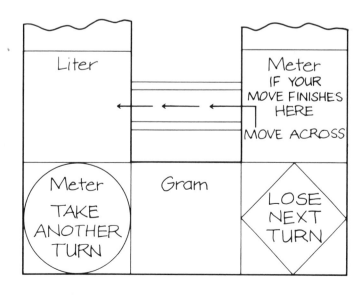

If METER is written in the first space after the START space, and the sequence is continued GRAM, LITER, METER, GRAM, LITER, . . . (one word in each space), the last unit written in the FINISH space should be GRAM.

3. In addition to the metric unit, students write TAKE ANOTHER TURN in the circle blanks and LOSE NEXT TURN in the diamond blanks.

To make Purchase Cards:

1. At least 15 Purchase Cards are needed. Write the names of items on cards, only one item on a card. (Picture cards can be made instead of word cards by cutting out items from old magazines and catalogs and then pasting them on the cards.)

2. Here is a list of items you can use for your Purchase Cards. Please add any items that might be of special interest to your students—especially ethnic food products or items used by business and industry in your region.

(You'll notice we've indicated the metric unit used to measure each item. These units should *not* be written on the cards.)

METERS are used to measure these items:

- clothes line
- ribbon
- fabric
- rope
- fishing line
- string
- yarn
- upholstery fabric

GRAMS are used to measure these items:

- butter
- nails
- canned soup
- parsley
- cereal
- salt
- flour
- spices

LITERS are used to measure these items:

- apple juice
- oil
- bleach
- orange juice
- gasoline
- paint
- milk
- wallpaper paste

To make answer list (optional):

On a sheet of paper, list alphabetically each item that appears on the Purchase Cards and the metric unit used to measure it.

DIRECTIONS:

1. (two or three players) One player shuffles the deck of Purchase Cards and puts them in a pile face down near the game board.

2. The tallest person goes first. This person pulls the top card on the pile and shows it to the other players and announces whether the item is sold by the METER, GRAM, or LITER. (When this play is finished, the card will be placed at the bottom of the pile.)

3. If she is correct, she moves her playing piece to the next space that has that measurement written on it. If another player disagrees with the answer, she must state what she thinks is the correct answer. If she is correct, she gets to move her playing piece ahead to the appropriate space.

4. The first person to reach the FINISH line is the winner.

VARIATION:

If you would like to focus on only one type of unit—for example, units for measuring mass—fill in the game board with the words **MILLIGRAM, GRAM, KILOGRAM, METRIC TON.**

The following is a list of items you can use for Purchase Cards for this variation. The metric unit should *not* be written on the cards.

MILLIGRAMS are used to measure these items:
- aspirin tablets
- gelatin powder
- multivitamins
- vitamin C tablets

KILOGRAMS are used to measure these items:
- beef
- potatoes
- cheese
- rice
- dog food
- turkey
- ham
- watermelons
- lemons

METRIC TONS are used to measure these items:
- airplanes
- steam shovels
- elephants
- trucks
- ships
- whales

GRAMS—see list in game DIRECTIONS above.

Quick and Easy

A. MEASURE YOUR GUESS

1. (Everyone in the class closes their eyes.)

2. (One person is chosen and asked to measure something in the classroom that is shaped like a rectangle; triangle; circle; cube; cylinder; and so on. She gives the shape and one of the measurements: "The rectangle is 3 feet long"; "The diameter of the cylinder is 1 centimeter"; and so on.)

3. (Everyone guesses what the object is. If no one names it, another dimension is given. Then other clues can be given such as color, texture, or where it is in the room.)

4. (The student who correctly names the item can be the next person to measure an item and have the class guess what it is.)

B. WHAT SHALL I WEAR?

1. (You–or a student–say, "The temperature is _____. What shall I wear?")

2. (Or, play it in reverse: "I'm wearing _____. What's the temperature?")

Examples:

Celsius	Fahrenheit	
–20	–10	long johns, lined boots, face mask
–10	10	heavy coat, boots
0	32	winter coat, hat, mittens
+10	55	light coat
+20	74	street clothes
+30	85	bathing suit

C. METRIC MERCHANDISE MEASURE

1. (Name a metric unit and ask students to name items that are typically measured in that unit.)

2. (On another day, reverse it. Name an item and ask what metric unit is used to measure it.)

Examples:

m — fabric, distances in the neighborhood
cm — paper, map scale
mm — nuts and bolts; sewing instructions
km — highway distances

Chapter 2: Measurement

sq cm	—	paper, desk tops
sq m	—	carpeting, playground areas
sq km	—	townships, ranches
g	—	spices, cereal, sugar
kg	—	butter, cheese, meat, potatoes, people
mL	—	medicine
L	—	milk, juice

D. SUPERMARKET SIZES

1. (State a specific measurement quantity: for example, 100 grams, 500 milliliters, 2 yards, 1 gallon.)

2. (Ask students to name something from the supermarket that is approximately that size. Encourage many responses to each quantity.)

Examples:

Customary Units	Metric Units	(*Note:* Quantities below are *not* equivalent)
1½ ounces	35 g —	cinnamon, basil
7 ounces	200 g —	candy bar, vanilla flavoring
1 pound	½ kg —	coffee, butter, cereal
2 pounds	1 kg —	honey, spaghetti
5 pounds	2½ kg —	flour, brown sugar
6 ounces	180 mL —	frozen juice concentrate
12 ounces	350 mL —	soft drink (can)
1 quart	1 L —	milk, mayonnaise
5 yards	5 m —	shelf paper, seam binding
200 yards	200 m —	thread, fishing line

E. CARPET FOR YOUR SECRET HIDEAWAY

1. If you had the best possible secret hideaway, where would it be? How big would it be? (List the places and their estimated size on the board.)

Examples:

closet	—	1 m x 2 m
tree house	—	2 m x 3 m
under the floor	—	3 m x 8 m
tunnel	—	1 yd x 7 yd
attic	—	4 yd x 9 yd
rooftop	—	9 ft x 12 ft

2. Assume one square meter (yard) of carpet costs $20. How much will it cost to carpet the floor in your secret hideaway?

F. PAINT A WALL

1. (Write on the board):

   Red Paint — 1 gallon covers 300 sq ft
   (1 liter covers 30 sq m)

   Yellow Paint — 1 gallon covers 400 sq ft
   (1 liter covers 50 sq m)

   Pistachio Paint — 1 gallon covers 600 sq ft
   (1 liter covers 75 sq m)

2. How many gallons of red paint will it take to cover 1200 sq ft of wall? 2400 sq ft? 3000 sq ft? 150 sq m? 300 sq m?

3. How many gallons of yellow paint? pistachio paint?

G. HELPING OTHERS METRICALLY

1. (Everyone's adoption of the metric system will bring the people of the world closer together. In that spirit, we offer this last Quick and Easy idea.)

2. Have the students ask their families, neighbors, and friends to donate food items to help the needy. Ask for items labeled with metric measurement—canned food, cake mixes, dairy products, and so on. (Many communities have a Sharing Shelf, Ecumenical Hunger Project, or similar program that would enthusiastically welcome food donations from a school group.)

Geometry

The Shape of the News   Activity 3-2

1. *Geometry for the Art Gallery*
2. *The Shape of the News*
3. *A Boxing Lesson*
4. *Bigger, Bigger, Smaller, Smaller*
5. *Stamp-Out Game*
6. *Tessellation Tuesday: Filling Space with Shapes*

7. *Go Fly a Kite!*
8. *Quick and Easy*
   A. *Make Room for Geometry*
   B. *Shapes of Signs*
   C. *Specialized Gift Stores*
   D. *Teased by Tessellations*
   E. *Parallel and Perpendicular*

Looking at real-world geometry is a great way to get a new perspective on the world around you. Why are foods packaged in cylinders and rectangular solids? Why are kites usually built in a diamond or box shape? Why is pie shaped in a circle and cake in a square?

This chapter is designed to answer these and many more questions and to help students take a closer look at the geometry around them. Using magazines and newspapers, tiles, stamps, boxes, and kites, students examine the practical as well as the aesthetic qualities of geometric shapes.

It will be a simple task to locate the materials you'll need for these activities—dot and graph paper are on pages 282, 283, and 284; magazines and newspapers are available for the asking (see sample letter, page 257); and geometric shapes and solids are everywhere.

We hope this chapter will help you and your students enjoy a new look at the world through geometric glasses.

Geometry for the Art Gallery

Students use pictures of geometric shapes to make collages, or use three-dimensional shapes to build geometric structures.

Math Skills: recognizing geometric shapes • geometric vocabulary

Curricular Areas: art

Materials: old magazines or catalogs • scissors • paste • poster paper • (optional: geometric solids such as cardboard tubes, boxes, and spools; string)

These geometric collages and sculptures are not only attractive decorations, they're also a delightful way for your students to get acquainted with the shapes of geometry.

DIRECTIONS:

1. (Divide the class into small groups.) Please look through your magazines and catalogs for pictures that show specific geometric shapes. Let's list some of those things and their shapes. (wheel—circle; tabletop—rectangle; checkerboard—square; sail—triangle; can of soup—cylinder; orange—sphere; tent—tetrahedron; ice cream cone—cone; ice cube—cube; cereal box—rectangular solid; and so on.)

2. I'll give the name of one shape to each group. In your magazines or catalogs, look for a variety of pictures that illustrate your shape. Find at least six. Arrange them on your paper to make a collage.

VARIATION: Have students collect items that are the shape of geometric solids (rectangular solid, tetrahedron, cone, cylinder, sphere, and so on). They can use these to make three-dimensional sculptures or mobiles.

The Shape of the News

Students find and outline geometric shapes in the newspaper. (*See photo, page 49.*)

Math Skills:	recognizing geometric shapes • geometric vocabulary (optional: recognizing symmetry)
Curricular Areas:	reading • vocabulary • art
Materials:	newspapers • markers

Trademarks, logos, and pictures in the ads consist of many geometric shapes. This treasure hunt will challenge students to a careful search for these different shapes.

DISCUSSION:　What shape is a newspaper? What other geometric shapes can you find in the newspaper?

DIRECTIONS:

1. Let's make a list of the geometric shapes you might find in the newspaper. (List students' responses on the board.)

2. We're going to hunt for these geometric shapes in the newspaper. As you go through your paper and find these shapes, outline them with your markers.

3. See how many different geometric shapes you can find. Look for different sizes of those shapes, too. Please label each shape with its name. If you wish, you may cut out your shapes and paste them on a large piece of paper.

VARIATION:　Have students find and outline pictures or drawings that illustrate symmetry.

A Boxing Lesson

Students identify various configurations of squares that will fold into a box with a lid. They then design boxes of various sizes and shapes.

Math Skills:	working with geometric shapes and spatial relations
Curricular Areas:	art
Materials:	Activity Sheet 3-3 • 1-cm graph paper (page 284) • scissors

---

Many things we purchase are packaged in boxes. This activity presents the basic idea behind designing and manufacturing these boxes.

DISCUSSION: What item have you recently purchased that came in a box? Do you think the box was specially designed for this item? What small items can you think of that are often packaged in boxes that make these items look larger?

DIRECTIONS:

1. (Have students complete Activity Sheet 3-3.)

2. (After students complete the Activity Sheet) On your graph paper, outline six adjacent squares that could be cut out to make a box with a lid. (This is an exercise where students will make lots of false starts as they try to find a solution. Fortunately, this trial-and-error process is a very effective way to learn!)

3. Suppose you need to design a box for a long, skinny item. Use your graph paper to design a rectangular box. (Have students design boxes of different heights, widths, and lengths.)

A Boxing Lesson

Which of these figures can be folded into a box? Check your answer(s) by cutting and folding.

1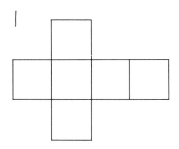

2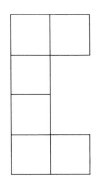

3

4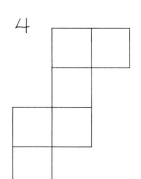

5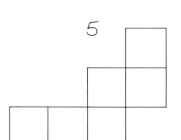

6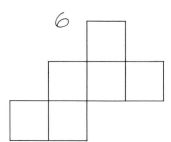

7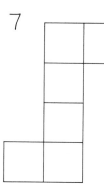

8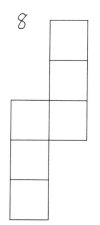

A BOXING LESSON

Bigger, Bigger, Smaller, Smaller

Students use graph paper to enlarge and reduce pictures and drawings.

Math Skills: working with coordinate geometry • proportional drawing

Curricular Areas: art

Materials: Activity Sheet 3-4 • 1-cm and 5-mm graph paper (pages 284 and 283)

One way to enlarge or reduce a drawing is to copy it on a grid and then transfer the drawing to a grid with larger or smaller squares. Here's an activity to practice this skill—and to lead the way for students to enlarge their own designs for book covers, T-shirts, posters, and other items.

PREPARATION: Duplicate materials above so that each student has an Activity Sheet, one sheet of 1-cm graph paper, and one-half sheet of 5-mm graph paper.

DIRECTIONS:

1. (Have students complete Activity Sheet 3-4.)

2. (After students complete the Activity Sheet) On a half-sheet of 5-mm paper, outline a figure. (Students could trace or draw a picture, or geometric shapes, or their initials using large, wide letters.)

3. Please label the bottom and left sides with letter and number coordinates like the ones on the Activity Sheet.

4. Exchange drawings with a friend. Using your 1-cm paper, make an enlargement of your friend's drawing. Be sure to turn the graph papers the same way. It's easiest if you start by placing several key parts of the drawing on the corresponding lines or positions.

VARIATIONS:

1. As a special project, some students might like to make or trace a drawing and then enlarge it to make a poster, a T-shirt, a book cover, or another item that can be used by the class or a friend.

2. Students could make up computer programs to enlarge and/or reduce designs.

Bigger, Bigger, Smaller, Smaller

Make a proportional drawing on the grid next to each picture.

Large Picture

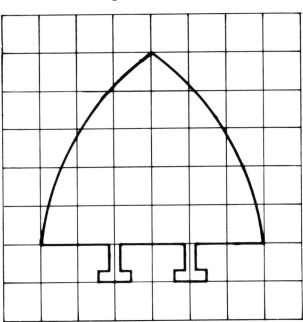

Small Picture

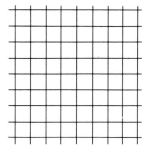

Small Picture

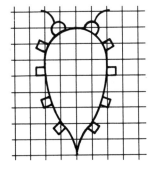

Large Picture

Stamp Out Game

Students make and play a game involving rectangles attached in different configurations.

Math Skills: working with geometric shapes and spatial relations • problem solving

Curricular Areas: art

Materials: (for each game): two sheets of 25¢ stamps (page 64) • two sheets of cardboard (8½ x 11) • small paper bag • scissors • (optional: clear contact paper)

A classroom set of these games takes some time to put together, but once you have them, . . . they'll stamp out boredom for years to come.

PREPARATION:

1. If you want to make a classroom set of games you'll need one game for every two or three students. Or you may want to let students work together to make and then play the game.

2. For each game, make two copies of the sheet of 25c stamps (page 64). Paste the stamps onto cardboard. Apply clear contact paper, if you want the game to last longer.

3. Using only one of the sheets of stamps, cut out all 25 shapes (pentomino pieces) shown on page 63.

4. The second sheet of stamps is the playing board. Using a marker, color the five stamps that form a cross in the center of the board. Cover the board with clear contact paper if you like.

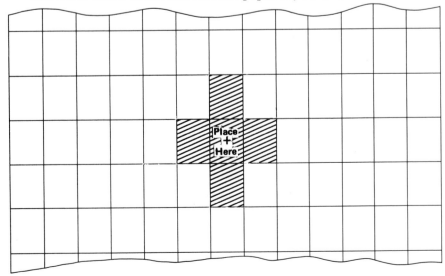

DIRECTIONS:

(*two to three players*)

1. Put the playing pieces in the paper bag. Players take turns pulling one piece at a time out of the bag until all the pieces have been distributed evenly. The person who pulls the cross piece should also take the last piece in the bag—giving him one more piece than the other players.

2. The player with the cross piece goes first by placing it on top of the colored cross on the playing board.

3. The next player places one of his pieces on top of other stamps on the playing board (but not on top of the other player's stamp piece).

4. Players continue to take turns placing their pieces on the playing board until it is impossible to fit in any more pieces. (If a player cannot fit any more of his pieces on the board, the opponent can continue to play any or all of his pieces as long as they do not overlap any pieces already played.)

5. The player with the least number of pieces left wins.

Stamp Out Game Pieces
(Use for Activity 3-5.)

Sheet of 25¢ Stamps
(Use for Activity 3-5 and Project 8-5.)

Tessellation Tuesday: Filling Space with Shapes

Students use dot paper to draw tessellations of various geometric shapes.

Math Skills: recognizing symmetry • working with geometric shapes

Curricular Areas: art

Materials: 2-cm dot paper (page 282) • (optional: markers or crayons)

Tessellation (or tiling) means using geometric shapes or other figures to completely cover a space. No gaps or overlaps allowed! Once students learn about tessellation (on Tuesday or any other day!), they'll see it everywhere—on wall and floor tiles, bricks, sidewalks, carpets, wallpaper, patchwork quilts, cabinet doors, air vents—and even on the crackers they munch.

PREPARATION: Make enough copies for each student to have at least six of the 2-cm dot paper on page 283. For simple tessellations, have students use a half sheet of dot paper.

DISCUSSION: Is it possible to fill a space (tessellate) with squares of the same size? What's a good example of a surface that's completely covered with squares? (classroom ceiling, sidewalk, bathroom wall, soda crackers, and so on) Do triangles (squares, hexagons) tessellate? (Yes) Do pentagons (octagons, circles) tessellate? (No)

DIRECTIONS: Squares: Ceiling Tiles

1. On your 2-cm dot paper, connect four dots to make a square.

2. Make another square right next to it so that the two squares have one side in common.

3. Continue making squares until you've filled your paper.

4. (Optional: Color the squares with different colors, using the same pattern throughout the tessellation.)

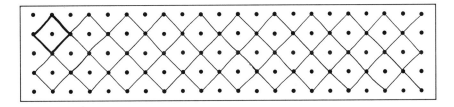

Chapter 3: Geometry

Rectangles: Bricks

1. Connect four dots to make a rectangle.

2. Continue making a pattern for a brick wall, filling the entire space with adjacent rectangles of the same size. Be sure to offset every other row of rectangles.

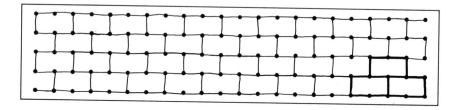

Triangles: Floor Tiles

1. Connect three dots to make a triangle.

2. Continue making a pattern for floor tiles with triangle tessellations. Triangles may be in any position as long as they are the same size and shape.

3. (Optional: Create a tessellation pattern by coloring adjacent triangles different colors.)

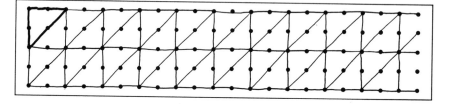

Trapezoids and Parallelograms: Carpets and Wall Paper

1. Connect four dots to form a trapezoid or parallelogram that you think will tessellate.

2. Fill the page with the shape you have chosen.

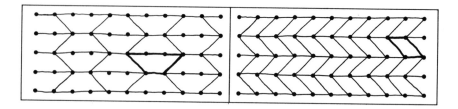

Octagons and Squares: Floor Tiles

1. Connect eight dots to make a regular octagon.

2. Make two more octagons, one next to it and one below it, so that each has one side in common with the first octagon.

3. Make a fourth octagon so that a square forms between four octagons. Fill the page in this way.

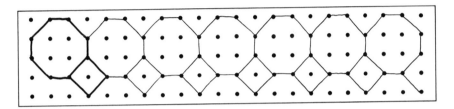

Two or More Geometric Shapes: Patchwork Quilt

1. Develop a pattern with two or more shapes that tessellate.

2. (Optional: Color the pattern to make an attractive design.)

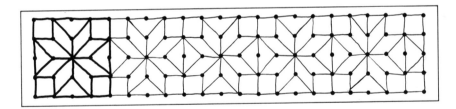

VARIATIONS:

1. Sometime when you're planning a snack, why not serve tessellating crackers? Buy several different kinds and shapes of crackers (squares, rectangles, circles, triangles, and so on) and let students experiment with tessellations while they enjoy their snack!

2. If students have access to LOGO on a computer, encourage them to experiment with computer tessellations (or tessellating computers??!!?).

Go Fly a Kite!

Students build a kite in the shape of a tetrahedron. *(See photo, page 243.)*

Math Skills: working with geometric shapes • geometric vocabulary

Curricular Areas: science • art • physical education

Materials: kite pattern (page 71) • oaktag • straws (7¾") • tissue paper • string (15-pound fishing line works best) • glue • scissors

---

If you can tie a knot . . . you can do this activity!

PREPARATION:

1. Trace the kite pattern (page 71) on folded oaktag; cut around the pattern and open it out.

2. Staple the oaktag to several thicknesses of tissue paper and cut with scissors or a paper cutter. (Each tetrahedron needs two sets of tissue paper cut in the shape of the pattern. Each kite needs four tetrahedrons.)

3. Cut string approximately 2 meters long for each tetrahedron.

DIRECTIONS:

1. (An easy way to do this activity is to have each student build one tetrahedron. Since each kite needs four tetrahedrons, a team of four students can combine efforts to build one kite.)

2. To make one tetrahedron, you'll need six straws. First feed the fishing line through three straws and form a triangle by tying a knot. Keep one end of the string as long as possible. How many straws would you use to make a square? A pentagon? An octagon?

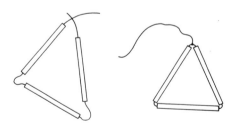

3. Feed the long end of the string through two more straws to make another triangle. Tie a knot. How many triangles are there? How many straws would you need to make two *separate* triangles?

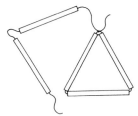 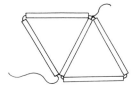

4. Feed the long end of the string back through one straw and attach the sixth straw to form a tetrahedron. How many triangular faces are there? How many edges? How many vertices? How many straws would it take to make four *separate* triangles?

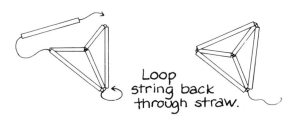

Loop string back through straw.

5. Lay one triangular face on the center of a tissue pattern. Fold flaps over and paste.

6. Paste another piece of tissue onto a second side. One of the flaps will overlap the first triangle. Does it matter which two sides you cover? Can you and three friends line up several tetrahedrons so that all the paper sides face exactly the same way?

7. (Four tetrahedrons are needed.) Line up the bases of three tetrahedrons. Make sure the covered sides all face *exactly* the same way. Tie them together.

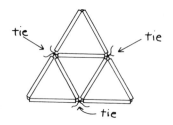

8. Tie a fourth tetrahedron on top. Again, make sure the two covered sides face the same way.

9. Attach a loose line of string along the edge where two covered sides meet. Tie the kite string to this loop.

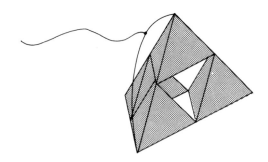

10. Go fly a kite!

VARIATION: Combine four kites to make a giant kite (16 tetrahedrons). It really will fly!

Kite Pattern
(Use for Activity 3-7.)

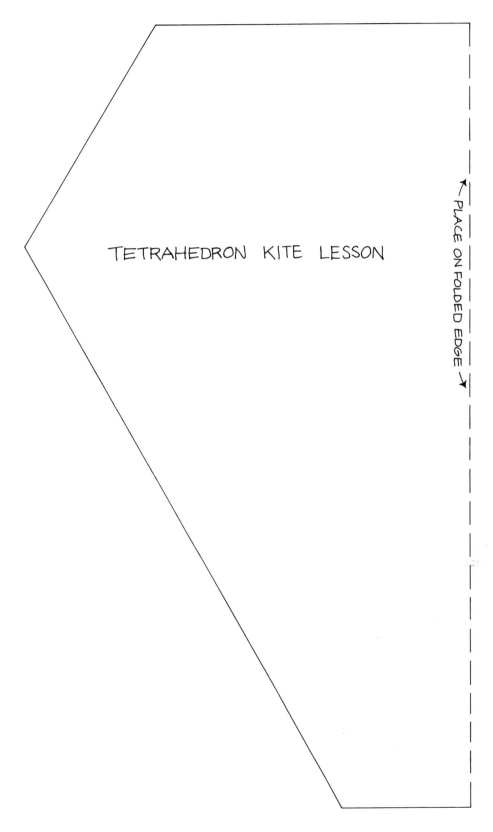

TETRAHEDRON KITE LESSON

← PLACE ON FOLDED EDGE →

Quick and Easy

A. MAKE ROOM FOR GEOMETRY

1. Close your eyes. Think about the things you see in your kitchen (or some other room). What do you see that is a rectangle? (circle, trapezoid, triangle, pentagon, hexagon, cylinder, sphere, cone, tetrahedron, rectangular solid)

2. (Use other places such as a department store, a candle shop, a dentist's office, or on a bus.)

B. SHAPES OF SIGNS

1. (On the board, draw a circle, triangle, and octagon.)

2. When you walk around town or ride in a car, what messages do you see on signs that are circles? (RAILROAD CROSSING) Signs that are triangles? (warning signs such as YIELD) Signs that are octagons? (STOP)

3. Now please draw some other traffic sign. Under your picture, write the name of its geometric shape.

C. SPECIALIZED GIFT STORES

1. You're going to open a gift store. You want to carry only sets of items that are congruent (same size and shape). What can you stock on the congruent shelves? (Sets of: dishes, flower pots, king-size sheets, bath towels, and so on.)

2. Your friends want to open a store that sells items that are similar (same shape). What might they have on the shelves for similar items? (small, medium, and large mixing bowls; three stack tables; beach ball and ping pong ball; bath and face towel; a set of cannisters; and so on)

D. TEASED BY TESSELLATIONS

(This is a good follow-up for Activity 3-6.)

1. On a piece of paper, please list all the tessellating surfaces you can see in this room. (for example, tiles, cupboard doors, window panes, air vents)

2. (Give students a chance to share their lists.)

3. (Have students list tessellating surfaces found in other places—sidewalk; floor, wall, ceiling tiles; patchwork quilt; and so on.)

E. PARALLEL AND PERPENDICULAR

1. If we looked for parallel lines at the football stadium (school

yard, construction site), where would we see them? (goal-posts, stairs, railings, metal frames, doors)

2. Now, let's look for perpendicular lines from the top of the World Trade Center (the post office, our school). Where can we see them? (TV antenna, lines of mortar in brick walls, window frames, climbing bars)

Part 2.
Application Activities

Earning and Banking

A Balancing Act   Activity 4-3

1. *Jingle, Pocket, Change Game*
2. *Checking Up!*
3. *A Balancing Act*
4. *The Lemonade Stand*
5. *Paper Route Math*
6. *A Budget—At My Age???*
7. *The $tate of $alarie$ Today*
8. *Help Wanted*

9. *Quick and Easy*
   - A. *A Penny for Your Thoughts*
   - B. *Check Your Money*
   - C. *Bill's Bills*
   - D. *This Taxes My Brain!*
   - E. *We Do Windows*
   - F. *Baby-Sitting—The Big Payoff!*
   - G. *Give-A-Hoot Recycling Center*
   - H. *Profit from Pickled Peppers*
   - I. *Life Income*
   - J. *Giant Salaries*
   - K. *Double Your Money!*

C H A P T E R 4

These activities are guaranteed to make sense (cents?!) for you and your students . . . or double your money back. (Play money, that is!) They are designed to help your students become better acquainted with earning and saving, as well as learn and review basic money skills.

Helping students learn about ways they can actually earn money is one important focus of this chapter. Other activities focus on how adults earn money—how they get jobs and how much they make. These activities can lead to interesting comparisons of jobs and incomes; they may also provide the spark for great dreams about how much your students will earn in their lifetimes.

Real-life experiences with the world of money can be provided through a field trip to a bank or a savings and loan company. (See page 259.) If field trips aren't in your budget this year, consider inviting a banker to come to school to explain and discuss the banking business. In addition to helping students learn, activities that bring your students and business people together are a good way to help the community keep in touch with the good job you and the school are doing.

Jingle, Pocket, Change Game

Students play this game in pairs. One person picks an amount of money less than $10 and the other person tries to guess the amount.

Math Skills: assigning place value • problem solving

This is a great game for all ages. Students can use simple or complex problem-solving strategies.

DIRECTIONS: (two players)

1. Decide who goes first. Pick an amount of money less than $10 and write it on a piece of paper so no one can see it.

2. Your opponent now has to guess that amount. She should write each guess on a piece of paper for you both to see.

3. Use the following system to let your opponent know how accurate a guess it was. Say "Change," if no digits are correct; "Jingle," if one digit is correct but in the wrong place; "Pocket," if one digit is correct and in the right place. If more than one digit is correct, say the appropriate word more than once.

Example:

Step	Chosen Amount	Opponent's Guess	Response	Meaning
1	$4.63	$1.92	"Change"	No correct digits
2		$1.42	"Jingle"	One digit correct but in the wrong place
3		$3.42	"Jingle, jingle"	Two digits correct, but in the wrong places
4		$4.21	"Pocket"	One digit correct and in the right place
5		$4.37	"Jingle, pocket"	Two digits correct but one is in the wrong place
6		$4.63	"Pocket, pocket, pocket"	You got it!

4. The play continues until the amount is guessed "Pocket, pocket, pocket!") The guesser is allotted a number of points equal to the number of steps it took to find the amount.

5. At the end of a given time period, the person with the lowest number of points wins.

Checking Up!

Students practice writing checks.

Math Skills: writing number words

Curricular Areas: handwriting • spelling

Materials: blank checks (page 81)

This activity definitely has a grown-up flavor to it, yet it's easy to do.

PREPARATION: Duplicate at least six checks (page 81) for each student. Draw two blank checks on the board or to project on an overhead projector. Write a list of three check amounts and payees, such as: $250.00—teacher; $47.52—supermarket; $16.85—local restaurant.

DISCUSSION: What is a checking account? What are some ways you can put money into (or take money out of) a checking account? (automatic deposit/withdrawal; writing a check)

DIRECTIONS:

1. On your sheet are three blank checks. What kinds of information do you find on a check? What must be written on a check?

2. Let's write our first check to someone you live with—we can be generous and make it for $100! What is today's date? Please fill in the date and number, then write the payee's name on the line marked PAY TO THE ORDER OF. (Demonstrate on the board.) What do we write after the dollar sign? How do we write the amount in words? Please sign your name legibly on the last line.

3. Let's do another one that includes cents. Will one of you show us how to write a check for $3.78 for the local newspaper? (Have a student demonstrate at the board.)

4. Now, you try it out. Please complete three more checks for the people and the amounts on the board.

VARIATIONS:

1. Have students cut out two sheets of blank checks and staple them together. (See page 86 for instructions on making them into a realistic-looking checkbook.) Have students write a check to pay an entry fee to math class each day.

2. Have students brainstorm a list of ways computers are used to manage checking accounts.

Blank Checks (Use for Activities 4-2, 4-3.)

_____

99-044
1990

————————————— 19 ———

Pay to the
order of _____ $ _____

_____ DOLLARS

CONSUMER NATIONAL BANK
321 Moneybags Road
Coolcash, California 91234

04 1990 0442 365 2004

_____

_____

99-044
1990

————————————— 19 ———

Pay to the
order of _____ $ _____

_____ DOLLARS

CONSUMER NATIONAL BANK
321 Moneybags Road
Coolcash, California 91234

04 1990 0442 365 2004

_____

_____

99-044
1990

————————————— 19 ———

Pay to the
order of _____ $ _____

_____ DOLLARS

CONSUMER NATIONAL BANK
321 Moneybags Road
Coolcash, California 91234

04 1990 0442 365 2004

_____

Hey!
Cut it out!

A Balancing Act

Students calculate bill totals, write checks, and determine the bank balance. (*See photo, page 77.*)

Math Skills:	addition • subtraction • multiplication • writing number words
Curricular Areas:	reading • handwriting • spelling
Materials:	blank checks (page 81) and check register (page 85) • (optional: calculators • checkbook cover and instructions–pages 86–88)

Calculating bills, writing checks, and balancing the check register may help students be more understanding of the scowls, moans, and short tempers that occur when adults are trying to balance their checking accounts.

PREPARATION:

1. Duplicate copies of checks and the check register (pages 81 and 85)—one copy of each page for each student. (Students can make them into realistic-looking checkbooks by following the instructions on pages 86 and 87.)

2. On the board or overhead, write at least five bills that students will pay. Include the quantity and price for each item; leave subtotals and totals blank for students to calculate.

Sample bills

SPORTS STORE	
	EACH
2 cans tennis balls	
1 volleyball	
3 jump ropes	

FRIENDLY FOODS	
	EACH
3 hamburgers	
2 cheeseburgers	
4 chili	
7 milk	
2 orange juice	

3. Determine a bank balance for students to use; it should be *less* than the sum of the five bills.

DISCUSSION: Why do people use checks? How do they keep track of their bank balances?

DIRECTIONS:

1. Before you can pay any bills, you'll need some money. Please write $_____ in the balance column of your check register.

2. On the board are five bills. Please calculate the total for each bill.

3. Now you are going to pay your bills. Unfortunately, you don't have enough money to pay them all. Please decide which four bills you can pay that will bring your balance as close to zero as possible without overdrawing your account.

4. After you've decided which ones to pay, please write checks for them. Record each check and calculate your bank balance in your check register.

VARIATIONS:

1. Pay all five bills. How much are you overdrawn?

2. Another way to bring your bank balance as close to zero as possible is to change the quantities of some of the items on the bills. For example, order two hamburgers instead of three.

3. After students have brainstormed a list of ways computers are used to manage checking accounts, have them arrange the list in order from first to last step. Discuss the ways individuals can check on the accuracy of computerized statements and other records.

Check Register (Use for Activity 4-3.)

		CHECKS DRAWN OR DEPOSITS MADE	BALANCE ➤ FORWARD			
Check #	Date	TO	Deduct Check — Add Deposit +			
			Balance			
Check #	Date	TO	Deduct Check — Add Deposit +			
			Balance			
Check #	Date	TO	Deduct Check — Add Deposit +			
			Balance			
Check #	Date	TO	Deduct Check — Add Deposit +			
			Balance			
Check #	Date	TO	Deduct Check — Add Deposit +			
			Balance — — — — — — — FOLD			
Check #	Date	TO	Deduct Check — Add Deposit +			
			Balance			
Check #	Date	TO	Deduct Check — Add Deposit +			
			Balance			
Check #	Date	TO	Deduct Check — Add Deposit +			
			Balance			
Check #	Date	TO	Deduct Check — Add Deposit +			
			Balance			
Check #	Date	TO	Deduct Check — Add Deposit +			
			Balance			
			Less Total Fees If Any			
			Balance			

STEP 1

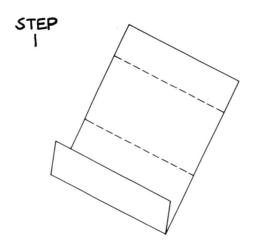

Fold your pattern
along the
dashed lines.

STEP 2

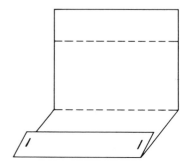

Fold bottom
up and
staple.

STEP 3

Cut out checks
and staple
them together.

STEP 4

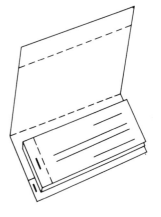

Staple the checks
to the bottom part
of the cover
at the left.

Chapter 4: Earning and Banking

Fold the top down.
Staple near the
edge.

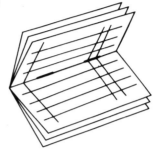

Staple together
2 or more
check register
pages. You must
staple at the
fold.

Slip the top
of the last
page of the
check register
into the
top fold.

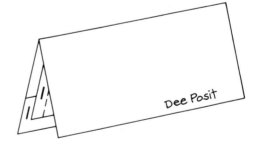

Dee Posit

Put your name on the
front cover.

Checkbook Cover (Use for Activity 4-3.)

- - - - - - - - - FOLD - - - - - - - - -

CUT LINE - - - - - - - - FOLD - - - - - - - -

- - - - - - - - - FOLD - - - - - - - -

The Lemonade Stand

Students calculate all costs involved in operating a lemonade stand, then estimate how much profit they can make.

Math Skills: addition • subtraction • multiplication

Curricular Areas: social studies • home economics

Materials: paper and pencils

Most of us have fond memories of our earliest entrepreneurial efforts —but were they really profitable ventures?

PREPARATION: Make a price chart as shown below; fill in approximate prices.

Paper cups (box of 50)	Lemonade mix (enough for 50 cups)	Water	Boards, nails, and paint	Advertising	Labor
$ _____	$ _____	$ _____	$ _____	$ _____	$ _____

DISCUSSION: In what ways do you earn money in your neighborhood? What are some of the expenses involved in setting up your own business?

DIRECTIONS:
1. Today we're going to calculate the cost and estimated profit for a lemonade stand.

2. On this chart, I've listed some materials and supplies that are needed for a lemonade stand. Can you think of anything I've left out? (Add students' suggestions to the chart.)

3. Please calculate the cost for operating the lemonade stand.

4. What should we charge for a cup of lemonade? (As students suggest prices, have them calculate total income for 50 cups and compare income with cost.) What would be the most realistic charge for a cup of lemonade?

5. (Optional: Have students be responsible for selling refreshments at a school function. Compare estimated costs and sales with actual costs and sales.)

VARIATION: Have students compare this lemonade-stand activity with one of the computer simulations about lemonade stands.

Paper Route Math

Students simulate the monthly accounting done by a local newspaper carrier. If someone in the class has a paper route, the class can help do her monthly report.

Math Skills: addition • subtraction • multiplication • division

Curricular Areas: reading • social studies

Materials: your local newspaper's rate schedule for their paper carriers • a blank accounting statement (page 91) • a completed statement (page 92)

This activity is bound to get the attention of anyone in your class who has a paper route—or anyone who knows someone who does!

PREPARATION: Find out if anyone in the class has a paper route or knows someone who does. Obtain the statement they filled out for the previous month's business (papers delivered and money collected). Make copies of both the completed statement and the blank statement for every student.

DISCUSSION: How do you think a paper carrier's pay is determined? If a customer doesn't pay, who loses—the paper carrier or the newspaper?

DIRECTIONS:
1. Each month, a paper carrier must balance the accounts for her delivery route. (If a student has a paper route, have that person describe a typical day, then share with the class last month's accounting. Otherwise, use the blank statement on page 91 and the completed version of the same statement on page 92.)

2. Let's look at this completed statement. (Discuss the information it contains.)

3. Here's a blank statement for each of you. (List on the board all the information they'll need for this month's accounting. Your paper-route students should be eager to provide you this information. Have the class balance the account.)

VARIATION: Have students develop a computer program (or select a commercially available one) to maintain records for a paper route.

Blank Accounting Statement for Paper Route
(Use for Activity 4-5.)

GRAVEL CITY GAZETTE

200 Rocky Road, Gravel City

_____ 19 _____

Account

For Period Ending _____

	Charge	Credit	Balance
_____ Daily average customers @ _____ per subscriber			
Carrier and Bond Insurance			
Misc. charge			
Credit for prepaids			
Credit for returns			
Credit for good service, 4¢ per customer			
Carriers may deduct 5¢ per customer discount if this statement is paid in full by the fifth at the bank.			

Date	Daily Draw
1.	
2.	
3.	
4.	
5.	
6.	
7.	
8.	
9.	
10.	
11.	
12.	
13.	
14.	
15.	
16.	
17.	
18.	
19.	
20.	
21.	
22.	
23.	
24.	
25.	
26.	
27.	
28.	
29.	
30.	
31.	
Total Papers	

Completed Accounting Statement for Paper Route
(Use for Activity 4-5.)

GRAVEL CITY GAZETTE

200 Rocky Road, Gravel City

Account _____

For Period Ending __December__ 19 ___

	Charge		Credit		Balance	
90 Daily average customers @ 3.85 per subscriber (round average to smaller number)	346	50				
Carrier and Bond Insurance	2	50				
Misc. charge rubber bands	1	75			350	75
Credit for prepaids			7	70		
Credit for returns			—			
Credit for good service, 4¢ per customer			3	60	339	45
Discount 5¢			4	50	334	95
					—	

90 × .05 = 4.50

Carriers may deduct 5¢ per customer discount if this statement is paid in full by the fifth at the bank.

Date	Daily Draw
1.	92
2.	
3.	
4.	
5.	—
6.	
7.	
8.	
9.	90
10.	
11.	
12.	
13.	—
14.	
15.	
16.	
17.	
18.	
19.	
20.	—
21.	89
22.	
23.	
24.	
25.	
26.	
27.	—
28.	
29.	
30.	
31.	89
Total Papers	2434

644

900

890

90+

Chapter 4: Earning and Banking

A Budget— At My Age???

Students plan a personal budget.

Math Skills:	estimation • addition • subtraction • (optional: working with percentages)
Curricular Areas:	handwriting • home economics
Materials:	paper, pencils • (optional: calculators)

After students have given some serious thought to their personal budget needs, they may want to share their proposed budgets with their parents.

DISCUSSION: What is a budget? Why do people have budgets? How can a budget be helpful to you?

DIRECTIONS:

1. What kinds of things do you buy with your money? (clothes, snacks, records, models, video games, gifts)

2. On a sheet of paper, list the kinds of things you have bought in the last week. (Help students select useful categories to describe their expenses.) Are there other categories that you would also need in a budget? Please add them to the list.

3. Beside each category, write the approximate amount you spent for such items in the past week. Look at these amounts and decide if that's a reasonable week's expenditure. If not, decide what the amount should be for an average week. Estimate an average week's expenses for the categories you listed.

4. What would be a reasonable amount to put in a savings account each week?

5. Please total your estimated weekly expenses. Would your total be a reasonable amount for you to spend each week? If not, what changes are needed to make it reasonable?

6. Show your budget to your parents. They will be impressed. They might even help you make your budget a reality!

VARIATIONS:

1. Ask students what percentage of their total budget was used for clothes; for entertainment. How many of them used more than 50 percent of their budget for food?

2. Ask students who have access to computer software for household budgets to share the programs. Have students compare the different programs and evaluate them in terms of their usefulness for keeping their own budgets.

The $tate of $alarie$ Today

Students compare salaries for various occupations in different parts of the country.

Math Skills: **reading and interpreting charts • addition • subtraction (optional: working with percentages • making graphs)**

Curricular Areas: **reading • social studies**

Materials: **Activity Sheet 4-7 • an almanac (optional: map)**

This activity is full of surprises. And they may or may not be happy ones—especially if you decide to compare your salary with that of a plumber . . . or of a government official . . . or of a teacher in another part of the country.

PREPARATION:
1. The annual salaries for various occupations in each of the 50 states are listed in most almanacs. Select two occupations for the students to compare.

2. On your first copy of Activity Sheet 4-7, write the job names as headings for the first two columns. (sample below)

3. Next to the abbreviation for each state, fill in the average annual salary for each occupation. (An aide or student helper could do this.)

State	Teachers	Plumbers	
AL	$35,000	$40,000	
AK	$70,000	$50,000	
AR			

4. Duplicate enough copies of filled-out Activity Sheet 4-7 for every student.

DISCUSSION:
Does a governor (lieutenant governor, office worker, teacher, doctor, auto mechanic) make a higher salary in one part of the country than another? Why do you think that's true?

DIRECTIONS:
1. Here's a chart (see PREPARATION above) showing the average annual salaries of two different occupations. What kinds of questions can this chart answer? (Within the same occupation, in what state do you find the highest/lowest salaries? Which place has the largest/smallest *difference* between salaries for the two occupations? And so on.)

2. What number could be filled in the blank column? (the difference between salaries; the percentage one salary is of the other; the salary earned in another occupation; and so on)

3. (Have students complete Activity Sheet 4-7.)

VARIATIONS:

1. Have students determine the best type of graph to use to present the information found in the chart, and then draw the graph.

2. Have students make a list of salaries from highest to lowest. You may want to let students write a computer program to put the salaries in order.

3. Give the students a map of the country. Have them fill in the salaries for each state. Ask them to look for relationships of salaries in different parts of the country.

4. (In this variation, students find salaries and write them in a blank chart.) Duplicate pages in an almanac. Have students choose an occupation, fill in the information on a chart, and compare salaries.

The $tate of $alarie$ Today

STATE							
AL				MT			
AK				NB			
AR				NC			
AZ				ND			
CA				NV			
CO				NH			
CT				NJ			
DE				NM			
FL				NY			
GA				OH			
HI				OK			
IA				OR			
ID				PA			
IL				RI			
IN				SC			
KS				SD			
KY				TN			
LA				TX			
MA				UT			
MD				VA			
ME				VT			
MI				WA			
MN				WI			
MS				WV			
MO				WY			

Help Wanted

Students search the classified ads to find and compare salaries being paid for different kinds of jobs.

Math Skills: subtraction • multiplication • comparison

Curricular Areas: reading • social studies

Materials: classified ads (employment section) for each student • Activity Sheet 4-8 • (optional: calculators)

The enormous differences in salaries given in the classified ads will spark considerable discussion about the fascinating world of jobs and salaries!

DISCUSSION: What does the word *salary* mean? How would you compare a job that earns $800 per month with one that earns $5 an hour? $250 a week with $13,000 a year?

DIRECTIONS: (Have students complete Activity Sheet 4-8.)

Help Wanted

(You will need one page of the employment section from the classified ads.)

1. Please read the employment ads. Find as many salaries as you can and circle each one.

2. In your ads, find the highest and lowest annual, monthly, weekly, and hourly salaries. Write them in the spaces below. Find the differences in each category.

		SALARY	JOB TITLE
	Highest	_____	_____
ANNUAL	Lowest	_____	_____
	Difference	_____	
	Highest	_____	_____
MONTHLY	Lowest	_____	_____
	Difference	_____	
	Highest	_____	_____
WEEKLY	Lowest	_____	_____
	Difference	_____	
	Highest	_____	_____
HOURLY	Lowest	_____	_____
	Difference	_____	

3. Using the HOURLY salaries from the first page, calculate the earnings for an 8-hour day and a 40-hour week.

	Hour	8-hour day	40-hour week
Highest HOURLY salary	_____	_____	_____
Lowest HOURLY salary	_____	_____	_____

Compare these salaries with the WEEKLY salaries.

4. Use the monthly salaries to calculate the earnings for one year (12 months).

	Month	Year
Highest MONTHLY salary	_____	_____
Lowest MONTHLY salary	_____	_____

Compare these salaries with the ANNUAL salaries.

Quick and Easy

A. A PENNY FOR YOUR THOUGHTS

1. (Put some coins in a box or can and shake it.)

2. (Students determine the coins and amounts by asking questions with "yes" or "no" answers:

   "Are the coins worth less than a dollar?" "Do you have any quarters in there?" "Do you have three dimes and a penny?")

3. (Write "1¢" on the board to denote a penny for each question asked. When the answer is found, find the total value. Students try to keep the total as low as possible.)

B. CHECK YOUR MONEY

1. (On the board, make a chart with five columns. Label them 50¢, 25¢, 10¢, 5¢, 1¢ going from left to right. Students make the same chart on lined paper at their desks.)

2. Put checks in the appropriate columns to show which coins you can use to make 60¢.

3. How many different ways did we find? Which solution used three coins? Which solution used the most coins?

4. (Have students keep their charts available for other moments when you want to do this Quick & Easy activity again.)

C. BILL'S BILLS

1. Mr. Bill Phold withdrew $100 from his savings account. The teller gave him exactly 50 bills. How many $1, $5, $10, and $20 bills did Bill receive? (possible answer: 40–$1; 8–$5; 2–$10.)

2. (Have students make up similar problems.)

D. THIS TAXES MY BRAIN!

1. I bought a ZINGO-STAR computer game for $21.95. If the sales tax was 5 percent, how much was the tax? ($1.10) How much did I have to pay altogether? ($23.05)

2. (Have students make up problems about other purchases and sales tax.)

3. (Have students use the 5 Percent Tax Table (page 107) to solve similar problems.)

E. WE DO WINDOWS

1. (Have students brainstorm jobs they can do in the neighborhood and list them on the board. Then have students decide

how much they could charge for each job—for example, $1.25 per hour; $.30 per window; $.75 for the first dog, $.50 for each additional dog)

Some possible jobs:

- wash windows
- water plants or feed pets for neighbors on vacation
- walk a dog
- baby-sit
- rent-a-party (plan games, refreshments, prizes for a birthday party)
- do someone's Christmas shopping for them
- collect paper, bottles, cans for recycling
- wash cars
- shovel snow
- mow lawns
- clean houses
- paint house numbers on the front curb
- distribute advertising flyers for a local business
- distribute campaign literature

2. (Have each student pick one job and write a word problem about how much money she would earn doing that job.)

F. BABY-SITTING—THE BIG PAYOFF!

1. (Write typical baby-sitting fees on the board. For example:

   $1.00 per hour

   $1.25 per hour after midnight

   $ .50 extra per child over two children)

2. (Have students take turns describing baby-sitting jobs, including the hours and number of children. Others calculate how much the job pays—for example, three children from 9 p.m. to 2 a.m. would pay $5.50.)

G. GIVE-A-HOOT RECYCLING CENTER

1. The Give-a-Hoot Recycling Center pays 30¢ per pound for aluminum cans. How much will George get for 5 pounds? 10 pounds? 25 pounds?

2. The rate for newspapers is 2¢ per pound. How much will Carolyn get for 10 pounds of papers? 25 pounds? 100 pounds?

Copyright © 1986 by Addison-Wesley Publishing Company, Inc.

H. PROFIT FROM PICKLED PEPPERS

1. I purchased 10 pickled peppers for 20¢ each and sold them for 30¢ each. How much did I make?

2. Paula paid $22.40 for four packs of pickled peppers and sold them for $8 per pack. How much was her profit?

3. Peter Piper purchased a peck of pickled peppers for. . . .

I. LIFE INCOME

1. (Quick and easy)

   (Give an example such as: "Lucy Long earns $20,000 a year. If she gets that much each year for 10 years, how much will she earn? How much would that be for 40 years?")

2. (Not so quick or easy!)

   If Lucy received a 10 percent raise each year, what would her 10-year income be? 40-year?

3. (Have students solve similar problems starting with monthly and weekly salaries.)

J. GIANT SALARIES

1. (Ask students if they have heard of anyone whose annual salary is more than $100,000—for instance, entertainers, TV personalities, baseball and football players.)

2. (Get them started by naming a fictitious person and salary. Have students calculate the income for five years; ten years. Go the other way: find the income per month, week, hour.)

3. (Encourage students to use an almanac and the daily news to find more giant salaries.)

K. DOUBLE YOUR MONEY!

1. (Riddle) What is the best way to double your money?

2. (Answer) Fold it!

5 PERCENT TAX TABLE

TRANSACTION	TAX	TRANSACTION	TAX
.01 – .10	.00	.90 – 1.09	.05
.11 – .27	.01	1.10 – 1.29	.06
.28 – .47	.02	1.30 – 1.49	.07
.48 – .68	.03	1.50 – 1.69	.08
.69 – .89	.04	1.70 – 1.89	.09
1.90 – 2.09	.10	2.90 – 3.09	.15
2.10 – 2.29	.11	3.10 – 3.29	.16
2.30 – 2.49	.12	3.30 – 3.49	.17
2.50 – 2.69	.13	3.50 – 3.69	.18
2.70 – 2.89	.14	3.70 – 3.89	.19
3.90 – 4.09	.20	4.90 – 5.09	.25
4.10 – 4.29	.21	5.10 – 5.29	.26
4.30 – 4.49	.22	5.30 – 5.49	.27
4.50 – 4.69	.23	5.50 – 5.69	.28
4.70 – 4.89	.24	5.70 – 5.89	.29
5.90 – 6.09	.30	6.90 – 7.09	.35
6.10 – 6.29	.31	7.10 – 7.29	.36
6.30 – 6.49	.32	7.30 – 7.49	.37
6.50 – 6.69	.33	7.50 – 7.69	.38
6.70 – 6.89	.34	7.70 – 7.89	.39
7.90 – 8.09	.40	8.90 – 9.09	.45
8.10 – 8.29	.41	9.10 – 9.29	.46
8.30 – 8.49	.42	9.30 – 9.49	.47
8.50 – 8.69	.43	9.50 – 9.69	.48
8.70 – 8.89	.44	9.70 – 9.89	.49
		9.90 – 10.09	.50

Shopping

The Hundred-Dollar Daydream   Activity 5-7

1. *What's My Change Game*

2. *Cash Register Tape Arithmetic*

3. *Inside an Invoice*

4. *Meat Labels Make Mean Math Lessons!*

5. *What's the Best Buy?*

6. *Same Item—Different Store*

7. *The Hundred-Dollar Daydream*

8. *Catchy Catalog Calculations*

9. *Down the Supermarket Aisle*

10. *The Shopping Center Game*

11. *Quick and Easy*
   A. *A Money-Go-Round*
   B. *But I Only Want One Pickle!*
   C. *Plenty of Postage Purchases*
   D. *Leisure-Time Tickets*
   E. *Birthday Money to Spend*
   F. *How Much Information Do You Need?*
   G. *How Math Helped Me Yesterday*
   H. *Quick Catalog Calculations*
   I. *Catalogs—A Weigh to Go!*
   J. *Are Computer Statements Always Accurate?*
   K. *Flowcharting A Shopping Spree*
   L. *Who Said Computers Rule Our Lives?*

Most children get their first real-world math experience in the seat of a shopping cart. They hear their parents make selections ("Let's buy a watermelon"), they experience measurement ("Let's see how much it weighs"), they learn about money ("Augh! That'll cost too much"), and they get a taste of solving problems ("Then let's get half of one . . .").

Because students know about shopping, it's easy for them to get started with the activities in this chapter. And once they get started, they'll learn and reinforce a multitude of math skills.

These activities are loaded with real-world money problems. They include practice in many different math skills as well as opportunities to use calculators and computers. But there's more to it than that. The activities also require making judgments, choosing wisely, defending choices—the thinking skills that are essential for critical consumers of all ages.

By the way, you might want to take a look at some of the shopping projects in Chapter 8. These projects, along with the activities found here, will be especially appealing to students because of the familiar settings. And as the students become wiser in their shopping skills, they will be eager to talk about what they're learning—a great opportunity for parents to notice the positive influence of school!

What's My Change Game

In groups of two to four, players take turns spinning an amount of a purchase. For each spin, they calculate the change they would receive.

Math Skills: **making change • addition • subtraction**

Materials: **game spinner (page 280) • paper clip, pencil • (optional: calculator)**

This game turns subtraction into fun. Play it for a few minutes or a whole hour—it will become a favorite activity all year long.

PREPARATION:

1. Decide on a starting amount from which students will make change. ($5, $20, $50, and so on.)

2. Duplicate game spinner (page 280). Divide each section in half. Write various amounts of money in the ten sections on the spinner to represent purchase amounts. Each amount must be less than the starting amount.

DIRECTIONS:

(two to four players)

1. Everyone spin a purchase amount and calculate your change. Who has the most change in your group? You'll go first when we start playing.

2. First player spins and calculates the change received from the starting amount.

3. Play continues to the left.

4. Each time a player spins, he calculates his change from his previous total.

5. Winner is the player whose money lasts the longest (or the first player to run out of money).

VARIATION:

To emphasize subtracting from 0: Each time a player spins, he calculates his change from his starting amount. Each player keeps a running total of all his change.

Cash Register Tape Arithmetic

Students practice addition and subtraction by finding the subtotals, totals, and differences (change) recorded on cash register tapes.

Math Skills:	addition • subtraction • estimation
Curricular Areas:	reading
Materials:	cash register tapes from stores • liquid correction fluid or small stickers • (optional: clear contact paper • calculators)

Kids need lots of practice adding and subtracting. This activity is perfect for taking the boredom out of that necessary drill.

PREPARATION:

1. Save all your cash register tapes from purchases made at various stores. Have your family, friends, and students save them too.

2. Separate tapes into groups according to the kind of arithmetic problems you want your students to practice. For example, make one group of tapes with two or three items purchased (simple addition), make another group with one-item purchases showing change received (subtraction), and so on.

3. If you just want students to add the amounts on each tape, use correction fluid or stickers to cover up the total. Cover the back of the tape if those numbers show through.

4. If you want students to find subtotals, totals, and change received, then cover these numbers.

5. Draw a rectangle to indicate where there once was an amount.

6. Code each group of tapes with a letter. Then number each tape. List the code letters and numbers on a separate piece of paper. Next to each number, write the answer the students should write in the blank rectangle on the corresponding tape.

7. (Optional: Cover the tapes with clear contact paper. They'll last for years this way. And students can write on them with non-permanent markers.)

DISCUSSION:

Why do stores give you the cash register tapes for your purchases? What do you do with them?

DIRECTIONS:

1. (Distribute a group of tapes to each student. Work with just a few students at a time until your collection of register tapes gets big enough for the whole class!)

2. List the code numbers of your register tapes along the left-hand side of your paper.

3. Estimate the amount missing on each tape and then calculate it exactly. Write the answer on your paper next to the appropriate code number.

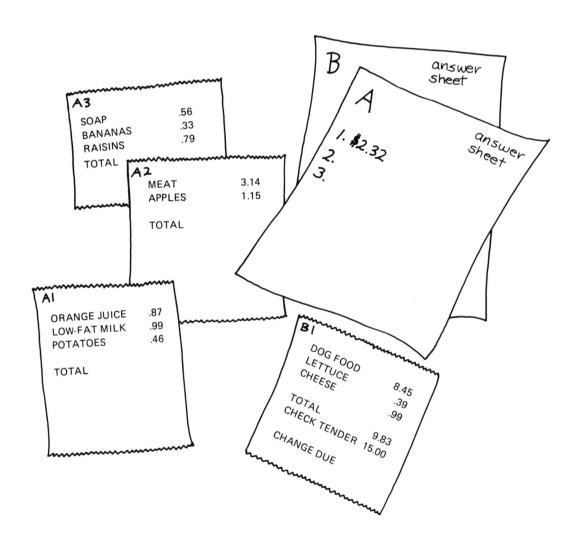

A3
SOAP   .56
BANANAS   .33
RAISINS   .79

TOTAL

A2
MEAT   3.14
APPLES   1.15

TOTAL

A1
ORANGE JUICE .87
LOW-FAT MILK .99
POTATOES   .46

TOTAL

B1
DOG FOOD
LETTUCE
CHEESE   8.45
             .39
TOTAL   .99
CHECK TENDER 9.83
             15.00
CHANGE DUE

B answer sheet

A
1. $2.32
2.
3.
answer sheet

VARIATIONS:

1. If you have calculators, this makes a good practice lesson. Use l-o-n-g tapes and have the students find sums and differences.

2. If you have computers, encourage your programming enthusiasts to write programs to calculate subtotals for nontaxable and taxable items, then calculate the tax and find the total bill.

Inside an Invoice

Using newspaper ads, students write invoices including all the essential information.

Math Skills:	addition • multiplication • estimation • (optional: making charts and tables)
Curricular Areas:	reading • handwriting • spelling
Materials:	invoices (page 115) for each student • local newspaper ads • staplers • (optional: calculators • 5 Percent Tax Table–page 107)

Although your students are probably too young to have jobs that require writing invoices, they'll enjoy a look INside an INvoice.

PREPARATION:

1. Duplicate two pages of invoices for each student.

2. List three items and prices from the Third Hand Store on the board: desk without drawers, $17.00; broken chair, $7.49; cracked mirror, $2.98.

DISCUSSION:

What is an invoice? (statement of amount due) What information is on the invoice? (date, items purchased, cost per item, total cost)

DIRECTIONS:

1. (Pass out two pages of invoices to each student.) Please look at the first invoice on your sheet. What information is needed on the invoice? (date, items, quantity, price for each, total price)

2. Let's make out the first invoice to show the purchase of one desk, one mirror, and three broken chairs from the Third Hand Store. What shall we fill in first? (store name and date) How do we fill in the items? (Have a student demonstrate at the board.) Quantity? Price for each? Total price? Check your answers by estimating the total.

3. (Distribute newspaper ads.) Please use the newspaper ads to find some merchandise you might like to buy. Fill out three more invoices, each with the name of a different store. Include at least three purchases on each invoice. One invoice should show that you purchased four of the same item (for example, four records; four shirts).

4. Now estimate the total price you owe to each store. Calculate the total and compare it with your estimate. You can staple each invoice to its newspaper ad when you turn them in.

VARIATION:

Use the 5 Percent Tax Table (page 107) to find the tax for each invoice. Add it into the total amount due.

Invoices (Use for Activity 5-3.)

INVOICE   DATE: _____

INVOICE   DATE: _____

Meat Labels Make Mean Math Lessons!

Students calculate the weight, the price per pound, or the cost of the package of items from the meat counter.

Math Skills:	multiplication • division • estimation
Curricular Areas:	reading • home economics
Materials:	labels from packages of meat, fish, poultry, and cheese • index cards • liquid correction fluid or small stickers • paste, tape, or staplers • (optional: calculators)

Beware! You may start serving some strange meals at home–just because the labels on the meat packages were perfect *for this lesson.*

PREPARATION:

1. Save labels that show price per pound, weight, and cost of package from your meat and cheese purchases. Have friends and students save them too. (This lesson will be especially real for students if the name of the local store appears on the label.)

2. Paste (tape or staple) each label on a small index card.

3. Make three arbitrary groups of cards and label each group with an A, B, or C. Then number the cards in each group.

4. On a separate piece of paper (which will serve as your answer sheet), list the code numbers (A1, A2, A3 . . .; B1, . . .). Next to the A codes, write the price per pound listed on each of the A cards. Next to the B codes, write the weights listed on each of the B labels. Next to the C codes, write the cost of the package listed on each of the C labels.

5. On the index cards, using liquid correction fluid or small stickers, cover the numbers you just recorded on your answer sheet. Be sure to cover only one number per label.

DISCUSSION: What information is given on the labels of meat or cheese packages?

DIRECTIONS: (Do this activity with a small group until you've collected enough tapes and cards for the whole class.)

1. Here are four meat labels with some information missing. For each label, decide whether to multiply or divide to find the missing number. Estimate the answer and then do the calculation.

2. Number your answer sheet with the code number on the card, and write your answer next to it.

VARIATION: Have students make up computer programs to calculate costs of meat and cheese.

Copyright © 1986 by Addison-Wesley Publishing Company, Inc.

What's the Best Buy?

Students calculate unit prices for different sizes of the same item.

Math Skills: division • comparison • making charts

Curricular Areas: reading • home economics

Materials: large newsprint paper • small, medium, large, and giant-sized packages for items such as toothpaste, cereal, and detergent (including containers of at least three different sizes for each item) • (optional: local newspaper ads • calculators)

Although it's important to know which size package has the lowest unit price, it's also important to understand that price is not always the deciding factor in choosing the best buy!

PREPARATION: Plan to divide the class into small groups. (Students can work individually if there are enough containers.) Draw a sample chart on the board:

Item:			
Size	Quantity	Price	Unit Price
Large			
Medium			
Small			

DISCUSSION: Why are items like cereal, toothpaste, and detergent sold in different-sized containers? Which sizes give the best buy for your money? What does "unit price" mean?

DIRECTIONS:

1. We'll do this activity in small groups. Will one person from each group come to the collection of boxes and pick up three different-sized containers of the same product (for instance, three Brand X cereal boxes, three Brand Y toothpaste boxes).

2. Find the price and the quantity on each package. Each person should calculate the unit price for each item. Check your solutions for accuracy.

3. Next, please draw a chart like the one on the board. Put the size of the container in the first column. If the size is named on the package (giant, super, large, and so on) use those names on the chart; if the size is not named, label them small, medium, and large.

4. Now, put a square around the lowest unit price, a triangle around the highest, and a circle around the middle price.

5. You can put your finished charts on the bulletin board where everyone can see them. Which size package usually had the lowest unit price? The highest? Is it always wise to buy the size with the lowest unit price? To buy the largest size? Why or why not?

VARIATION: Using local newspaper ads, have students make up best-buy problems such as:

A. Jan's Grocery: Juice—2 liters for $1.79

Mary's Market: Juice—1 liter for $ .89

Which is the best buy?

B. 12 oz octopus tentacles—$2.29

1 lb octopus tentacles—$2.59

Which is least expensive?

Same Item— Different Store

Students visit or phone different stores to find the price for a given item, then make graphs to compare prices among stores.

Math Skills: comparing costs • subtraction • making graphs

Curricular Areas: oral expression • social studies

Prerequisite: familiarity with bar graphs

Materials: 1-cm graph paper (page 284) • crayons or markers

You and your colleagues may benefit from the results of these price comparisons as much as the students will!

DISCUSSION: What are all the different kinds of stores where you can buy a record album? A hair dryer? A microcomputer game? A fishing rod? A board game? (department store; discount store; music store; gift shop; and so on) Which one usually has the lowest prices?

DIRECTIONS:

1. We're going to check on the prices for certain items in several stores. Then we'll make graphs to show price comparisons.

2. We'll work in teams of three or four. Each team will select one specific item; be sure to agree on the brand of the item you're going to price. Each person will be responsible for finding out the price for the item in one store (by phone or in person).

3. (Assign students to teams and have each team decide what specific item they will price. List the items and brand names on the board.)

4. (After the prices have been collected:) What kinds of stores did you call or visit? (List them on the board. Then identify a color for each kind of store; for instance, discount store—red; gift shop—yellow.)

5. Each team will make a graph to show the prices at the different stores. Mark the dollar amounts in a column on the left side. Use intervals that are appropriate to the price (for example, $5, $10, $15. . . ; $5.00, $5.10, $5.20).

6. For each store, fill in the graph by coloring in the price with the color listed on the board. At the bottom of each colored bar, write the name of the store. Please print the name of the item below the graph.

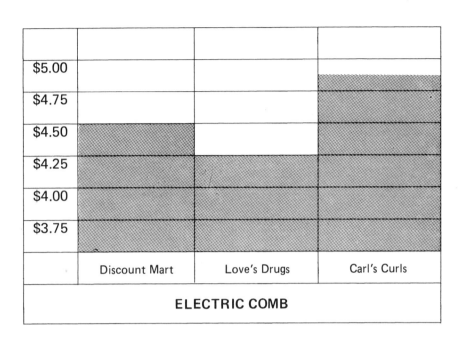

	Discount Mart	Love's Drugs	Carl's Curls
$5.00			
$4.75			
$4.50			
$4.25			
$4.00			
$3.75			

ELECTRIC COMB

7. (When graphs are completed:) Which kind of store had the lowest price for the record album? The hair dryer? (and so on) Did one kind of store consistently have the lower prices? The higher prices? Why do you think this is so?

VARIATION: Interested students could make up programs to create their graphs on the computer.

The Hundred-Dollar Daydream

Using a catalog or weekly shopping ads, students order four items that have a total price as close to $100 as possible. (*See photo, page 109.*)

Math Skills: addition • subtraction • estimation

Curricular Areas: reading • handwriting

Materials: catalog (or section of one) or weekly shopping ads for each student

---

Sometimes it just makes sense to let kids daydream—and learn at the same time.

DISCUSSION: Do you ever order items from mail order catalogs? What kinds of things are you (or your family) most likely to buy by mail?

DIRECTIONS:
1. Take time to look through your catalog pages or ads and see what kinds of things you might like to have for yourself and your family.
2. You want to spend $100. Please find four items with a total cost as close to $100 as possible.
3. On your paper, write each item and its cost. Calculate the total.

VARIATIONS:
1. Change the number of items students are to buy.
2. Change the amount of money given to each student.

Catchy Catalog Calculations

Students calculate the cost for several different items as well as for more than one of the same item.

Math Skills:	addition • multiplication • working with percentages
Curricular Areas:	reading • spelling • handwriting
Materials:	catalog page (page 41) • order form (page 42) • (optional: calculators)

Plan to use this activity for several math periods—it needs to go slowly. You know your students, so plan this activity to fit their pace.

PREPARATION: Duplicate the catalog page and the order form for each student.

DISCUSSION: Do you ever order items from mail order catalogs? What steps and calculations are involved in finding the total cost of a mail order?

DIRECTIONS: Lesson 1 (Adding prices)

1. Please look at your catalog page and select four items you'd like to buy.

2. Let's look at the order form. We'll only fill in part of the information today. Please find these columns on your form: Name of Item, Price Each, and Total Price.

3. Now, write the name of your first item in the Name of Item column and put its price in the column for Price Each. What will we write in the Total Price column?

4. Please fill in your next three items and their prices.

5. Then calculate the total for your items and write it in the space for Total for Goods.

Lesson 2 (Multiplying prices)

1. Today we're going to do another mail order. This time we'll order more than one of each item.

2. Please find three things you'd like to buy. Fill in the Name of Item and the Price Each as you did before.

3. In the column called How Many, please write two for the first item, three for the second one, and five for the third.

4. How do we calculate prices when we buy more than one? On scratch paper, please multiply the price of your first item by two, then write the amount in the Total Price Column on the same line. Multiply to find the Total Price for the other two items.

Lesson 3 (Multiplying and adding prices)

(If you're getting tired of that same old catalog page, this might be a good time to use pages from real catalogs. Just tear out some pages from a catalog and pass them out.)

1. Please look at your catalog page again and find five items to order. Write them in the Name of Item column.

2. Please order one of the first item, two of the second, three of the third, and so on.

3. Multiply the price of each by the quantity you're ordering, and write that amount in the Total Price column.

4. Please find the total cost of the items and write it in the Total for Goods space.

5. (Checking your students' work will be very time-consuming so you may want to have them check their own work by using a calculator. Or you may want to have them exchange papers and check each other's work.)

Lesson 4 (Finding percentages and adding prices)

1. Today we're going to do an order form and include the extra charges. On your catalog page, please find just one item to order.

2. Fill in the Name of Item, Price Each, Total Price, and Total for Goods.

3. Our first extra charge will be a 5 percent sales tax. Please calculate 5 percent of your total and write it in the Tax space. (Or you could distribute copies of the 5 Percent Tax Table—page 107.)

4. Now write $1.25 in the Shipping Costs space.

5. Add the Tax and Shipping Costs to the Total for Goods. Write that amount in the Total space.

Down the Supermarket Aisle

A picture showing products and prices in a section of the super-market is presented to the students. They use the picture to make shopping lists, find the cost of items, and calculate the total cost of their purchases.

Math Skills: addition • multiplication • (optional: subtraction • division • estimation • vocabulary • measurement • working with geometry)

Curricular Areas: reading • spelling • handwriting • home economics

Materials: grocery section (page 129) • produce section (page 130) • Activity Sheet 5-9 • (optional: calculators)

Two lifelike drawings of parts of a supermarket add realism to this activity. Students can color them or you can make them into trans-parencies and use them on the overhead projector. Presto! Your class is an instant supermarket.

PREPARATION:

1. Make one or both supermarket drawings (pages 129 and 130) into a transparency, and beg, borrow, or buy (?!) an overhead projector.

   If you don't have a projector, or you'd rather duplicate the pictures so each student can have them on paper, you can make one copy, fill in prices, then make multiple copies. [Be sure to indicate either pounds (lb) or kilograms (kg) for oranges, apples, onions, and potatoes on the produce price signs.]

2. For the adventurous teacher: Borrow two overhead projectors and project both drawings side by side. Presto! You've got a stereo supermarket in your classroom.

DISCUSSION: What sections of the supermarket are these? Let's fill in the prices on these signs. What amount would you suggest for these apples? the lettuce?

DIRECTIONS:

1. (Have students complete Activity Sheet 5-9.)

2. (Optional: Change the quantities on the Activity Sheet. Use correction fluid to white them out, then fill in new amounts.)

VARIATIONS: Use the drawings of the grocery section and the produce section for math practice in many different ways. Students can find:

- the price of more than one of the same item   (multiplication)

- the price of one item when the price
  is given for two   (division)
- change received from a purchase   (subtraction)
- the total price of many items   (addition—use a calculator)
- a total of a combination of prices offered
  in a variety of ways   (combined operations)
- approximate cost of a purchase   (estimation)
- price of a dozen, gross, and so on   (vocabulary)
- total volume, weight, length   (measurement)
- shapes   (geometry)
- and on and on   (etc!)

Down the Supermarket Aisle

Look at the pictures of the grocery section and the produce section. Fill in the prices for the items on your grocery lists.

1. (Use with grocery section)

1 box of noodles	_____
1 kg honey	_____
1 box green crystal soap	_____
1 can meat dog food	_____
1 bag of bagels	_____
TOTAL	_____

2. (Use with grocery section)

3 cans spinach juice	_____
6 bags of banana chips	_____
4 boxes waffle mix	_____
2 100 g honey	_____
3 jars peanut butter	_____
4 cans turkey dog food	_____
TOTAL	_____

3. (Use with produce section)

1 _____ potatoes	_____
1 dozen kumquats	_____
1 bunch carrots	_____
1 _____ apples	_____
1 head lettuce	_____
TOTAL	_____

4. (Use with produce section)

6 grapefruit	_____
5 _____ potatoes	_____
10 _____ apples	_____
3 _____ onions	_____
6 bunches carrots	_____
3 heads lettuce	_____
TOTAL	_____

Chapter 5: Shopping

The Shopping Center Game

Students design a game board with the names of local stores and items sold there. Players then move around the board buying items and subtracting the price from their bank balances.

Math Skills:	addition • subtraction
Curricular Areas:	spelling • handwriting • art
Materials:	game board (page 282) • game spinner (page 280) • paper clip and pencil • playing piece for each student (a coin, ring, or bean) • tally sheet (notebook paper) • (optional: calculators)

Kids love this game because they design it, build it, and play it—and because it's all about the shopping centers they know!

PREPARATION:

To make spinner:

1. Duplicate game spinner (page 280).

2. Fill in the five sections with the numbers 0, 1, 2, 3, 4.

To make game board:

3. Duplicate a game board (page 282) for each student. (Larger boards can be made by making a transparency and tracing around a projected image on tag board or stiffening material.)

4. Students fill in circle blanks with TAKE ANOTHER TURN and diamond blanks with FREE SPACE.

5. They fill each of the other 14 blank spaces with the name of a local business, an item or service sold there, and its cost.

To make player's tally sheet:

6. Find the total of the 14 prices and take half of it. Each player starts with this bank balance. Players write this balance at the top of their tally sheets. (Or use play money instead of a tally sheet.)

DIRECTIONS:

1. Players spin and the highest number goes first. This person spins again and moves his piece the given number of spaces.

2. Player records the amount of the item on his tally sheet and subtracts it from his balance.

3. Play continues to player's left. For each turn, the player spins and moves. If he lands on a business space, he subtracts the amount shown; otherwise, he follows the directions on the space.

4. All players play until they reach the finish line. (Those who run out of money before anyone reaches the finish line continue to subtract their expenses—keeping track of their negative balances.) The winner is the player with the highest bank balance.

VARIATION:

The winner is the person who's farthest in the hole!

Quick and Easy

A. A MONEY-GO-ROUND

1. If a soccer ball costs $14.48, how much is that to the nearest ten cents? to the nearest dollar? to the nearest five dollars?

2. (Have someone name another item and price, and have students round that amount to the nearest ten cents, five cents, dollar, and so on.)

B. BUT I ONLY WANT ONE PICKLE!

1. Today's special for pickles is two for _____ cents. What will one cost?

2. If a jar of pickles is marked "3 for _____ cents," what's the unit price?

3. If a 10-ounce jar costs _____ cents, how much is that per ounce?

4. (State additional items and prices for a given amount, volume, or weight. Have students figure unit price.)

C. PLENTY OF POSTAGE PURCHASES

1. If you had $2, how many _____¢ stamps could you buy? How many postcards?

2. (Ask the question in reverse. Name the quantity and price of stamps and have students find the total price.)

D. LEISURE-TIME TICKETS

1. Roller coaster tickets are 60¢ each. A book of six tickets costs $3. How much do I save on six rides if I buy the book?

2. Tickets for the soccer game are $3.50. Eight of us went to a game. How much did it cost altogether?

3. (Make up a word problem about buying tickets for a basketball game. For a Ferris wheel. For a movie.)

E. BIRTHDAY MONEY TO SPEND

1. (Write three to five items with their prices on the board. For example: roller skates—$30; radio—$20; wallet—$10; guitar—$40; computer game—$15.)

2. (You or your students make up questions for the class to answer: "Aunt Helen sent me $70 for my birthday. Which three items can I buy?" "My birthday gifts included three $10 bills and two $5s. What can I buy?")

F. HOW MUCH INFORMATION DO YOU NEED?

1. (Ask the class questions that relate to shopping at the supermarket. Tell them not to solve the problem—but just tell if you've given them the right amount of information to solve it.)

2. I bought bread, milk, and eggs, and paid $3.14. How much did the bread cost? (too little information)

3. I bought 3 kilograms of peaches for $1.50. How much does 1 kilogram cost? (just enough information)

4. Cereal costs 75¢, crackers cost $1.25, and cough drops cost 95¢. How much does it cost for two boxes of crackers? (too much information)

G. HOW MATH HELPED ME YESTERDAY

1. (Students tell or write a situation they experienced yesterday where they used math.)

2. (In each case they must include:
   a. Where—at Hotdog Heaven
   b. What math—addition
   c. What happened—I totaled my money to see if I had enough to buy the Super Dawg and Super Fries.)

H. QUICK CATALOG CALCULATIONS

1. (Pick up a catalog and open it at random. Slowly name three items and their prices. Have students mentally calculate the total price.)

2. (Gradually increase the number of items and prices.)

I. CATALOGS—A WEIGH TO GO!

1. (Pick up a catalog that includes shipping weights. On the board write the names of two to four items from this catalog. Next to each write the shipping weight.)

2. What's the total weight for these items?

3. (Point to the first item.) How much would two of these weigh? Two of the next?

4. Find the total weight of the first two items. Of two of the first and three of the next.

J. ARE COMPUTER STATEMENTS ALWAYS ACCURATE?

1. What kinds of information do you and your family receive on computerized statements? (utility bills, bank statements, charge account bills, cash register tapes, and so on)

2. How do you know if these statements are accurate?

3. (Have students discuss ways to check for accuracy. Discuss what should be done when there's an error.)

K. FLOWCHARTING A SHOPPING SPREE

1. (In mixed-up order, list on the board the steps you follow when you want to give your friend a birthday gift: wrap the package, look around in your favorite store, give the gift to your friend, pay the cashier, select a gift.)

2. (Ask students to rearrange these steps in their logical order and draw a flowchart to depict them.)

3. (Have students flowchart some amusing events such as making purple ice cubes. In mixed order, the steps are: close the freezer, put tray in freezer, get a tray, put purple food coloring in water, open the freezer, fill tray with water.)

L. WHO SAID COMPUTERS RULE OUR LIVES?

1. Many devices that affect our lives are partially or completely run by computers. Help me list them on the board. (automatic ticket sales, scoreboards, microwave ovens, traffic lights, thermostats, automatic banking, clocks, and so on.)

2. Let's classify the different uses we have listed.

Eating

Menu Math   Activity 6-1

Spaghetti, chicken, ice cream—who can resist a chance to talk about their favorite foods, their favorite places to eat?

This chapter is crammed with tasty activities. The first one, Menu Math, is actually a large collection of activities from very easy to very challenging. What grade level do you teach? You'll find plenty of activities here that are just right for your students. The chapter includes a game, lessons using order forms, a chance to take a careful look at pizza, and even a graph you can grow . . . and eat!

Except for food (!), everything you'll need for this "eating" chapter is here. You'll find menus and order forms for two different restaurants—any one of them can provide as much or as little math practice as you need. If you want to add extra flavor, it's fun to use menus from local restaurants. (We've included some ideas in Chapter 10 for collecting these menus.) As a special event, you might even order pizza for a snack or take the class on a field trip to a fast-food restaurant.

Students will feast on this smorgasbord of activities. In the process, they'll digest some important math skills and get a healthy taste of real-world problem-solving experiences.

Menu Math

Using menus and order forms, students do a variety of menu-related math problems. (*See photo, page 137.*)

Math Skills: numbers and numeration • estimation • comparison • addition • subtraction • multiplication • division • working with percentages

Curricular Areas: reading • spelling • handwriting • vocabulary • art

Materials: menu (page 143) • order forms (page 144) • (optional: 5 Percent Tax Table—page 107) • calculators • take-out or placemat menus from local restaurants—See page 272 for other ideas.)

Many minutes of meaningful mathematics are mixed up in menu math. As starters, here is an ample supply of ideas for students to do alone, in groups, or with the whole class.

PREPARATION: Make a copy of the menu and fill in appropriate prices. Duplicate menus and order forms for each student.

DIRECTIONS:

VERY EASY ACTIVITIES . . . To whet the appetite

After you've looked over your menus, we'll answer some questions. (After ten minutes or so, ask questions such as the following.)

1. How much does milk cost? What salad costs $2.75?

2. Find the highest/lowest-priced dinner (beverage, dessert . . .) on your menu.

3. Which sandwich costs about $3? about $4.50?

4. If you had $5, which dinner could you buy?

5. Which costs more, spaghetti or a hot roast beef sandwich?

EASY ACTIVITIES . . . With added spice

Look at your order forms. Take a few minutes to make up a name for the restaurant, fill it in at the top of the order forms, and draw a design if you like. (Present one or more of the following problems for students to solve.)

1. Using your order form, write a bill for one person and total it. Order a dinner, dessert, and beverage (or a sandwich and beverage . . .).

2. Estimate the cost for a chef's salad, milk, and yogurt. Then fill out an order form and compare the total with your estimate.

3. If you had $8, which dinner and dessert could you buy?

4. What change would I receive from a $10 bill if I ordered a hamburger plate?

5. I ordered a dinner and something to drink. My bill totaled $ _____ (price of one dinner and drink). What did I order?

TOUGHER ACTIVITIES . . . You can sink your teeth into

In these activities, we'll be ordering meals for more than one person.

1. Using your order form, write a bill for two (three, four . . .) people and total it. Order dinner, dessert, and beverage for each person.

2. Estimate the cost for two ham and cheese sandwiches and two shakes.

3. What change would I receive from a $20 bill if I ordered a vegetarian garden plate, juice, and fresh fruit? (Optional: Write an order form and show the change as well as the total.)

4. My three friends and I each ordered the same kind of sandwich and milk. Our bill totaled $ _____ (price of three sandwich and milk orders). What kind of sandwiches did we order?

5. Ask four of your classmates to place an order with you. Write out an order form for each friend. Total the bills.

HARD ACTIVITIES . . . For complete satisfaction

Now that you've had experience with menus and order forms, let's try some more complicated problems.

1. Find a 15 percent (10 percent, 20 percent . . .) tip on an order of one steak dinner. On another dinner, on a salad. (Optional: Round to the nearest 5 cents.)

2. Using your order form, write a bill for one or more people and total it. Add on 5 percent (6 percent . . .) tax.

3. Harry ordered spaghetti and a dish of ice cream. Carrie ordered a chef's salad and milk. Larry had a hamburger plate, milk, and ice cream. Sharie ordered spaghetti, milk, and a fresh fruit dessert.

   Using only one order form, record the above. The same items, even though ordered by different people, should be recorded on one line only. Total the order. Then calculate how much each person owes. (Optional: Calculate the tip each person should leave—10 percent, 15 percent, 20 percent.)

4. Let's practice splitting the bill. (Divide the class into groups of three, four, or five students. One student takes an order from the others. That student computes the bill and finds the totals. The other students then check it for accuracy. They figure out the amount each person owes and the tip each student should leave.)

5. Design your own menu. Be sure to include item names, descriptive words when appropriate, and prices. If you like, choose a theme, such as a breakfast menu, an ethnic menu (Italian, Mexican, Chinese, and so on), or an ideal school lunch menu!

6. Make up a computer program to

   A. Add the total bill.

   B. Multiply when two or more people order the same item, then add the total bill.

   C. Add the total food cost, calculate 5 percent (6 percent . . .) tax, then add the total bill.

   D. Calculate a 15 percent (20 percent . . .) tip.

   E. Add the total food cost, calculate the tax, calculate the tip, then add the total bill.

HAMBURGER PLATE

Beef patty on toasted. $.
bun with tomato, lettuce,
onion, french fries, and
salad
with cheese $.

SANDWICHES
all served with potatoes or salad

FRENCH DIP
sliced roast beef on a roll $.
served with our special gravy

HOT ROAST BEEF $.
HOT TURKEY $.

REUBEN
Sliced corned beef, swiss $.
cheese, and sauerkraut
on rye
HAM AND CHEESE $.
CHEESE $.
HAM $.

BEVERAGES

Milk . Iced tea .
Juice . Soft drinks .
Coffee . Small .
Tea . Large .
shakes .

SALADS
CHEF'S SALAD $.
Ham turkey, cheese,
tomatoes, and hard
boiled egg slices on
crisp greens

FRUIT SALAD $.
Fresh fruit in season

DINNER SPECIALS
Ask about
the special of the day!

DINNERS

COUNTRY FRIED CHICKEN $.
Includes salad, rolls, potatoes

VEGETABLE GARDEN PLATE $.
Fresh vegetable assortment

FISH 'N' CHIPS $.
Deep-fried cod with
french fries

SPAGHETTI $.

ROAST BEEF $.
Served with potatoes, rolls,
choice of vegetables or salad

STEAK—New York or Bullseye $.
Includes salad, rolls, and
potatoes

DESSERTS

Yogurt . Ice Cream .
Fresh fruit . dish .
Pie . sundae .

Order Forms (Use for Activities 6-1, 6-7.)

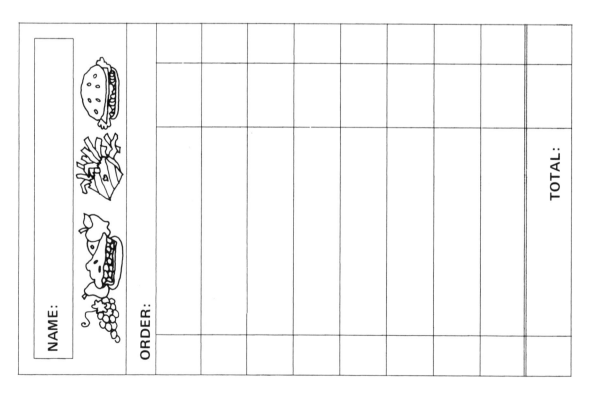

Fast-Food Finances

In three separate activities, students (1) use order forms to calculate the cost of fast-food purchases, (2) estimate costs of food orders, and (3) make a set of order forms.

Math Skills: estimation • addition • multiplication • making charts and tables • (optional: working with percentages)

Curricular Areas: reading

Materials: fast-food order forms (pages 147 and 148) • price signs (pages 149 and 150) • (optional: 5 Percent Tax Table—page 107 • clear contact paper • calculators)

Say the word hamburger *to your students and you'll have their attention immediately. Capitalize on this interest in fast food with these three activities.*

PREPARATION:

Activity 1:

Duplicate fast-food order forms (page 147). (To make permanent reusable forms, laminate or cover with clear contact paper.)

Activity 2:

Make an overhead transparency of the price sign on page 150 and project it on a wall or screen to simulate a sign at a fast-food restaurant. With the help of your students, fill in reasonable prices for each item. (If an overhead projector is not readily available, you can duplicate this page for everyone, or simply write the items and prices on the board.)

Activity 3:

Duplicate blank order forms (page 148). Fill in prices on price sign (page 150) and duplicate.

DISCUSSION: Why do fast-food restaurants take your order on special forms?

DIRECTIONS: Activity 1 (Forms with prices—page 147)

1. Let's pretend we work at a fast-food restaurant. Use one of the order forms to record the following order: one hamburger, one vegetable burger, two fries, one milk, one orange juice, two ice creams.

2. Circle the price in the appropriate column and record that amount in the Total column.

3. Find the total for this order.

4. (Optional: Have students find tax from the tax table on page 107 —or from a current one. They can record the tax and the grand total on the form.)

Activity 2 (No forms—just price sign—page 149)

1. Let's pretend we're going to eat at a fast-food restaurant. Estimate the cost of a hamburger, fries, milk, and ice cream. Of two small salads, a shake, and pie.

2. You only have $5. What can you buy?

3. (Have students make up additional problems to estimate.)

Activity 3 (Forms without prices—page 148)

1. (This is a good calculator adventure.) Let's pretend we own a fast-food restaurant. We need to make up new order forms. First fill in an appropriate price for one of each item on the form.

2. Multiply to find the price of two of each item, three of each, and so on.

3. (Optional: On the second form, increase each price by 10 percent.)

VARIATION:

Have students make up computer programs to do any of the above activities.

Fast-Food Order Forms (Use for Activity 6-2.)

Form 1 — 137952

NAME:

No.	Item	1	2	3	4	5	Total
	P'Burg	.80	1.60	2.40	3.20	4.00	
	C'Burg	.95	1.90	2.85	3.80	4.75	
	V'Burg	.50	1.00	1.50	2.00	2.50	
	Pit-Pat	1.75	3.50	5.25	7.00	8.75	
	Salad – Sml	.60	1.20	1.80	2.40	3.00	
	Salad – Pit	1.50	3.00	4.50	6.00	7.50	
	Fries	.55	1.10	1.65	2.20	2.75	

No.	Item	1	2	3	4	5	
	Shakes v c s	1.00	2.00	3.00	4.00	5.00	
	Juice o t g	.40	.80	1.20	1.60	2.00	
	Milk	.45	.90	1.35	1.80	2.25	
	Lemonade	.65	1.30	1.95	2.60	3.25	
	Ice Cream	.35	.70	1.05	1.40	1.75	
	Pie w wo	.75	1.50	2.25	3.00	3.75	
	Fruit	.30	.60	.90	1.20	1.50	

☐ ketc ☐ must ☐ onion   Eat Here ☐ To Go ☐

TOTAL

Pitts Burgers   137952

Form 2 — 137952

NAME:

No.	Item	1	2	3	4	5	Total
	P'Burg	.80	1.60	2.40	3.20	4.00	
	C'Burg	.95	1.90	2.85	3.80	4.75	
	V'Burg	.50	1.00	1.50	2.00	2.50	
	Pit-Pat	1.75	3.50	5.25	7.00	8.75	
	Salad – Sml	.60	1.20	1.80	2.40	3.00	
	Salad – Pit	1.50	3.00	4.50	6.00	7.50	
	Fries	.55	1.10	1.65	2.20	2.75	

No.	Item	1	2	3	4	5	
	Shakes v c s	1.00	2.00	3.00	4.00	5.00	
	Juice o t g	.40	.80	1.20	1.60	2.00	
	Milk	.45	.90	1.35	1.80	2.25	
	Lemonade	.65	1.30	1.95	2.60	3.25	
	Ice Cream	.35	.70	1.05	1.40	1.75	
	Pie w wo	.75	1.50	2.25	3.00	3.75	
	Fruit	.30	.60	.90	1.20	1.50	

☐ ketc ☐ must ☐ onion   Eat Here ☐ To Go ☐

TOTAL

Pitts Burgers   137952

Blank Fast-Food Order Forms (Use for Activity 6-2.)

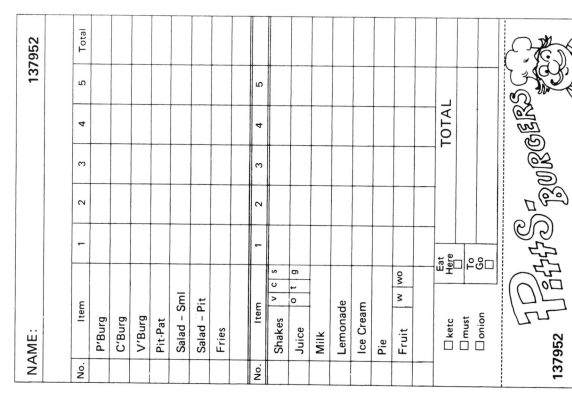

Price Sign (Use for Activity 6-2.)

PITTS-BURGERS
MR. PITTS LOW PRICES
PIT-BURGER $ .80
BARBEQUED HAMBURGER
CHEESE BURGER $ .95
FROM OUR BARBEQUE PIT
VEGETABLE BURGER $ .50
TRY ONE TODAY
THE BIG PITTS-PATTIES $ 1.75
2 PATTIES, LETTUCE, TOMATOES, SPECIAL SAUCE
FRUIT $ .30
WITH OR WITHOUT PITS
PIE $ .75
APPLE, CHERRY, SPINACH
ICE CREAM $ .35
VANILLA, CHOCOLATE, SPINACH
TRY OUR DELICIOUS SALAD BAR!
THE SALAD PIT $ 1.50
SMALL SALAD $ .60
GOLDEN FRIES $ .55
PITLESS JUICE $ .40
ORANGE, TOMATO, SPINACH
DRINKS
SHAKES $ 1.00
MILK $ .45
LEMONADE $ .65

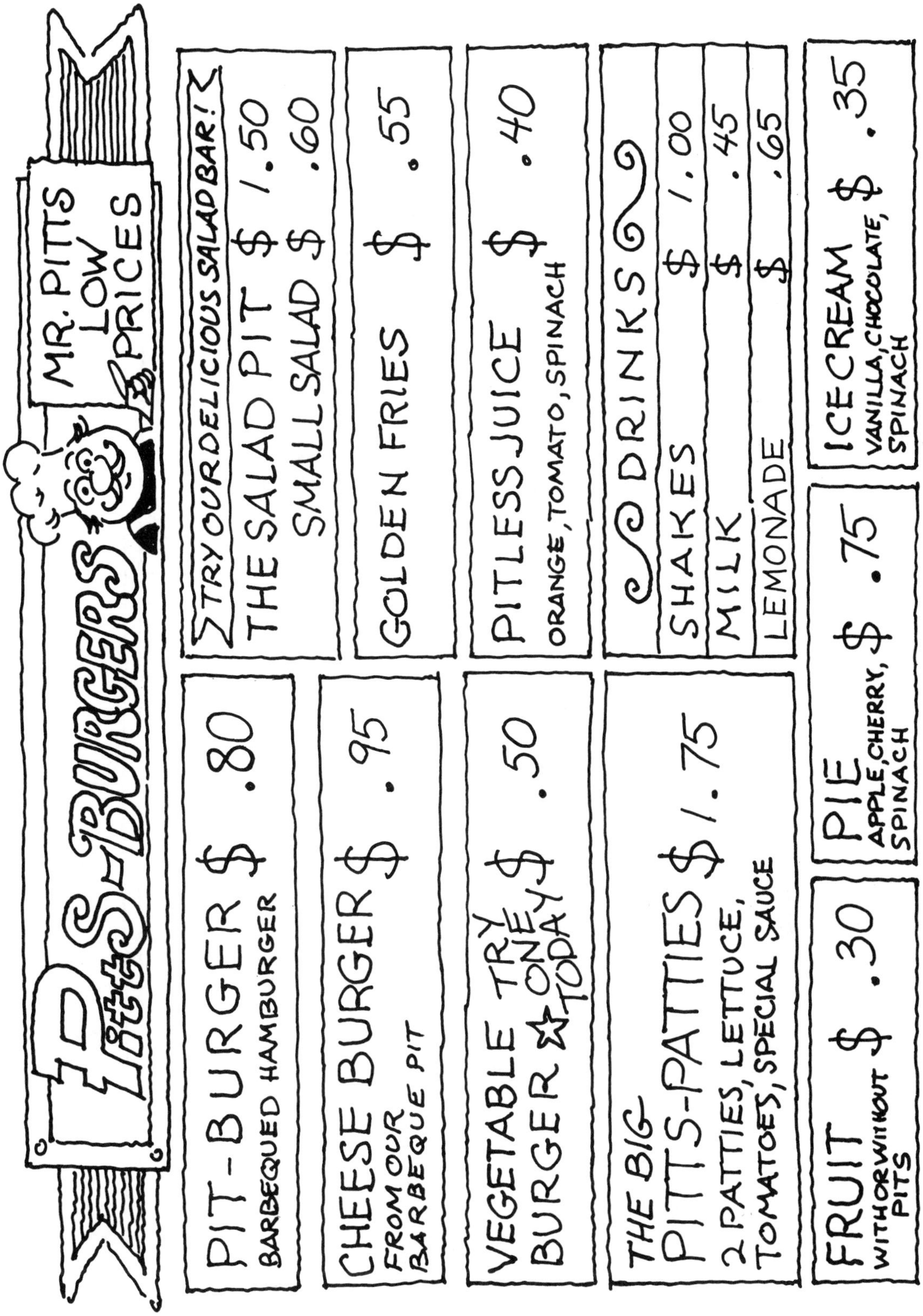

MR. PITTS
LOW PRICES

PITT'S BURGERS

→ TRY OUR DELICIOUS SALAD BAR!

THE SALAD PIT $ . .
SMALL SALAD $ . .

GOLDEN FRIES $.

PITLESS JUICE $.
ORANGE, TOMATO, SPINACH

∽ DRINKS ∽

SHAKES $ . .
MILK $ . .
LEMONADE $ . .

ICE CREAM $.
VANILLA, CHOCOLATE, SPINACH

PIT-BURGER $.
BARBEQUED HAMBURGER

CHEESE BURGER $.
FROM OUR BARBEQUE PIT

VEGETABLE BURGER $.
TRY ONE TODAY

THE BIG PITTS-PATTIES $.
2 PATTIES, LETTUCE, TOMATOES, SPECIAL SAUCE

FRUIT $.
WITH OR WITHOUT PITS

PIE $.
APPLE, CHERRY, SPINACH

A Mathematical Garden of Eatin'

Students will record the rate of growth of plant sprouts on a graph.

Math Skills: metric measurement—linear • making line graphs

Curricular Areas: science

Materials: 5-mm graph paper (page 285) • containers for plants (milk cartons, egg shells, egg cartons, ice cube trays, and so on • seeds (beans, grass, birdseed) • soil • centimeter rulers (page 281)

Plant sprouts are fun to grow . . . easy to measure and graph . . . and some are even quite tasty for a snack!

DISCUSSION: What kinds of plants do you have growing at home? Are they in pots, a window box, a garden? Why do people raise gardens and have plants in their homes?

DIRECTIONS:
1. We're going to plant some seeds, then observe how fast—or slow!—they grow. Will each of you please take a container, fill it with soil, choose some seeds to plant, then water it. You can put two or three bean seeds in one container; if you choose bird-seed, use eight to ten seeds; for grass seed, use a pinch.

2. Please label the container with your name and the kind of seed you've planted.

3. We'll observe and water our plants each day (not too much water, please). When the first sprout appears, we'll measure it and record its height on a graph.

4. On your graph paper, label the heights going up the left side— 0 mm, 5 mm, 10 mm, . . . 145 mm, 150 mm. Across the bottom, label the days 1 through 14. Day 1 will be the first day your sprout appears. We'll measure and record the heights daily for two weeks.

VARIATIONS:
1. Use carrot tops, sweet potatoes, or white potatoes. Measure the length of the roots as well as the vines.

2. Vary the way sprouts are grown by planting some seeds in soil and by putting some on a wet sponge or inside wet blotter paper.

3. Compare the heights of plants grown under different conditions. Using bean seeds planted in soil, put one pot in direct sunlight, one in indirect light, and one in a dark closet. Compare another group of plants by varying the amount of water used for each one.

Mathematical Popcorn

Students will compare the volume and mass of popcorn before and after it's popped.

Math Skills: **multiplication • division • measurement—volume and mass • (optional: comparing costs)**

Curricular Areas: **science • home economics**

Materials: **popcorn • oil • large containers • equipment for volume and mass measurement • one or more popcorn poppers (Use the sample letter on page 257 to ask parents to lend their popcorn poppers.)**

Here are some activities guaranteed to appeal to all five senses. You and your students will touch and measure popcorn, hear it expand, measure and see the increase in volume—and taste the results. (Needless to say—the smell will permeate the experiment.)

PREPARATION: Set up the popcorn popper in a safe place where it is visible to the class.

DISCUSSION: Why do you think popcorn pops when heated? (The moisture in the kernel expands.) How much more space do you think the popped corn will take than the unpopped corn? Will it weigh more or less?

DIRECTIONS:
1. Using one popper and the same brand of popcorn, we're going to make popcorn and compare the weight and volume of the corn before and after it's popped.

2. First, weigh these two empty containers. (Paper cups make good lightweight containers.)

3. Next, pour 25 mL (or 1 tbsp) of oil into one container, then weigh it; record its mass. Pour 100 mL (or ½ cup) of corn into the second container, then weigh it and record it.

4. Now pour the oil into the popcorn popper. Put two or three kernels of corn into the oil and cover the popper.

5. When the kernels pop, pour the rest of the corn in the popper and wait for the entire batch to pop.

6. Measure the volume of the popped corn. How does it compare with the unpopped measurements?

7. Weigh the popped corn. How do the before and after measurements compare? Record the results.

Chapter 6: Eating

8. Follow steps 2 through 7 for a second batch of the same brand of popcorn. How do the results of this batch compare? Record the results.

VARIATIONS:

1. (Compare different brands of popcorn.) Use one popper and two or three different brands of popcorn. Follow steps 2 through 8 above. Which brand produced the greatest volume of popped corn? The next greatest? How do their prices compare? Which brand is the best buy?

2. (Compare different brands of popcorn poppers.) Use several different poppers and one brand of popcorn. Follow steps 2 through 8 above. Which popper produced the greatest volume of popped corn? The next greatest? Which popper would you buy?

You Can Halve Your Cake and Eat It too!

Using a metric version and a customary version of the same recipe, students will determine the amount of ingredients needed for half the recipe, double the recipe, and ten times the recipe.

Math Skills: metric and customary measurement—volume • multiplication • division

Curricular Areas: science • home economics

Prerequisite: addition and multiplication of fractions

Materials: **Activity Sheet 6-5**

Without a word from you, your students will find out very quickly how easy *the metric system is—in contrast to its English cousin.*

PREPARATION: Write the following recipe for carob cake on the board. Copy both the metric and customary versions.

Carob Cake Recipe (customary measure)	**Carob Cake Recipe** (metric measure)
¾ c whole wheat flour	180 mL whole wheat flour
1 tsp baking powder	5 mL baking powder
¼ tsp salt	1 mL salt
½ c butter or margarine	120 mL butter or margarine
½ c carob powder	120 mL carob powder
⅓ c honey	80 mL honey
2 eggs	2 eggs
⅔ c sunflower seeds	160 mL sunflower seeds
3 tbsp milk	45 mL milk
¼ tsp vanilla	1 mL vanilla
Bake at 350° for 15-20 minutes.	Bake at 350° for 15-20 minutes.

DISCUSSION: When you double a recipe, how do you find amounts like two times 180 mL? Two times 2/3 cup? When you cut a recipe in half, how do you find half of 180 mL? Half of 3/4 cup? When you double the recipe, do you change the directions for mixing and cooking?

DIRECTIONS: (Have students complete Activity Sheet 6-5.)

VARIATIONS:
1. Have students bring their favorite recipes from home. They can exchange these recipes, then halve and double them.

2. See also A CLASS PARTY (Chapter 8, page 195).

You Can Halve Your Cake and Eat It too!

You will need two versions (customary and metric) of the carob cake recipe. Calculate the ingredients needed to change the recipes as described below.

CUSTOMARY MEASUREMENT

1. Double the recipe

_____ whole wheat flour
_____ baking powder
_____ salt
_____ butter or margarine
_____ carob powder
_____ honey
_____ eggs
_____ sunflower seeds
_____ milk
_____ vanilla

METRIC MEASUREMENT

1. Double the recipe

_____ whole wheat flour
_____ baking powder
_____ salt
_____ butter or margarine
_____ carob powder
_____ honey
_____ eggs
_____ sunflower seeds
_____ milk
_____ vanilla

CUSTOMARY MEASUREMENT

2. Cut recipe in half

_____ whole wheat flour
_____ baking powder
_____ salt
_____ butter or margarine
_____ carob powder
_____ honey
_____ eggs
_____ sunflower seeds
_____ milk
_____ vanilla

METRIC MEASUREMENT

2. Cut recipe in half

_____ whole wheat flour
_____ baking powder
_____ salt
_____ butter or margarine
_____ carob powder
_____ honey
_____ eggs
_____ sunflower seeds
_____ milk
_____ vanilla

Suppose your school is having a bake sale and you want to make ten of these cakes to sell. Calculate the ingredients you'll need.

Pizza—Price per Portion

Students calculate the cost of pizza by area.

Math Skills:	working with decimals • multiplication • division • calculating area • (optional: working with fractions)
Curricular Areas:	home economics
Prerequisite:	area of a circle • division of decimals
Materials:	Activity Sheet 6-6 • calculators

This activity gives calculator practice a special flavor!

PREPARATION: Before duplicating the Activity Sheet, fill in different prices for each of the six pizzas.

DISCUSSION: Is a small pizza the same amount as one-half of a large one? Which size pizza gives you the most for your money?

DIRECTIONS:
1. (If necessary, review the steps involved in finding the area of a circle.)
2. (Have students complete Activity Sheet 6-6.)

VARIATION: Have students make up computer programs to determine the area of a pizza and to calculate the cost per square centimeter (or inch). Compare the area cost of pizzas from several local pizza places.

Pizza—Price per Portion

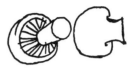

1. Calculate the area (A) of each pizza in square centimeters.
   ($A = \pi r^2$; $\pi = 3.14$)
   REMEMBER: The size of the pizza is its diameter. Divide the diameter in half to find its radius (r).

12 cm

EXAMPLE
Mini-Pizza

$3.39

Area: 113 cm²
Price per cm²:

$.03

$A = \pi r^2$
$A = 3.14 \times 6^2$
$\ = 3.14 \times 36$
$\ = 113 \text{ cm}^2$
$3.39 \div 113 = \$.03$

2. Figure the cost per square centimeter for each pizza.

VIC'S PIZZA PLACE

Small $ _____
20 cm

Medium $_____
30 cm

Large $_____
40cm

Area: _____cm²

Price per cm²:

_____

Area: _____cm²

Price per cm²:

_____

Area: _____cm²

Price per cm²:

_____

REGULAR PIZZAS

Esther's Eatery

$ _____

24 cm

Area: _____cm²

Price per cm²:

_____

H. D.'s Pizza

$ _____

24 cm

Area: _____cm²

Price per cm²:

_____

The Pizza Machine

$ _____

24 cm

Area: _____cm²

Price per cm²:

_____

Just for fun: Calculate the area and price per cm for pizza from your favorite pizza place.

The Restaurant Game

Students move around a game board that gives the names of food items found on a menu. They look up these items on a menu and record the foods and prices on an order form. The object of the game is to get the highest total on your restaurant bill!

Math Skills: addition • multiplication

Curricular Areas: reading

Materials: game board (page 282) • game spinner (page 280), paper clip, and pencil • playing pieces for each student (a coin, ring, or bean) • menu for each player (page 143) • order forms (page 144) • (optional: calculators)

Students of all ages enjoy this game. You can add extra appeal by using menus from a local restaurant.

PREPARATION: To make tally sheets:

Duplicate an order form (page 144) for each student.

To make menus:

Fill in appropriate prices on the menu (page 143). Duplicate one for each student.

To make spinner:

1. Duplicate game spinner on page 280.
2. Fill in the five sections with the numbers 0, 1, 2, 3, 4.

To make game board:

1. You or your students can fill in the following information on the game board. (If you plan to use menus from a local restaurant, use the following as a guide and fill in food items from the menu.)

   A. Fill in circle blanks with:
      - ORDER CHEAPEST BEVERAGE
      - ORDER CHEAPEST SANDWICH
      - ORDER CHEAPEST DINNER

   B. Fill in diamond blanks with:
      - ORDER MOST EXPENSIVE DESSERT
      - ORDER MOST EXPENSIVE SALAD
      - ORDER MOST EXPENSIVE DINNER

C. Fill in other 14 spaces with:
- 2 CHEF'S SALADS
- 3 ICE CREAM SUNDAES
- A HAM SANDWICH AND MILK
- 1 HAMBURGER PLATE
- 2 DIFFERENT DINNERS
- A HOT SANDWICH AND COLD DRINK
- 3 SPAGHETTI DINNERS
- 4 BEVERAGES (SAME)
- A CHEESEBURGER PLATE
- ANY SANDWICH AND DRINK
- 4 DESSERTS (SAME)
- 3 SALADS (SAME)
- A SANDWICH
- 2 ICED TEAS AND 3 SHAKES

2. Duplicate individual game boards (page 282) for all students. (Larger boards can be made by making a transparency and tracing around the projected game board on tag board or stiffening material.)

DIRECTIONS: (two to four players)

1. First player spins and moves the number of spaces indicated.

2. Player then looks up item(s) on the menu and records the name and price on her order form. If a player lands on a space that indicates two *different* items, each should be recorded on a separate line on the order form.

3. Play continues to player's left.

4. All players play until they reach the finish line. Players total their order forms. The winner is the player with the highest total. (Optional: The first player to reach the finish line adds a $2 tip to her total.)

VARIATION: The winner is the person with the lowest total.

Chapter 6: Eating

Quick and Easy

A. DINNER AT EIGHT

1. What time should I start fixing the potatoes if they take 20 minutes to peel, 40 minutes to cook, and 15 minutes to mash? (Dinner's at eight!)

2. When should I start fixing the rolls if they take 10 minutes to mix, 30 minutes to rise, 15 minutes to shape and put into the pan, and 25 minutes to bake?

B. CONCESSION STAND CALCULATIONS

1. (Write on the board three to five food items with prices similar to those at local concessions. For example, popcorn—75¢; hot dogs—$1.00; drinks: large—80¢, small—50¢; ice cream bars—90¢.)

2. (You and your students make up questions for the class to answer:)

   My friend and I ate lunch at the ball game on Saturday. We each ate a hot dog, popcorn, and a small drink. How much did we spend? I spent $1.50 at the swimming pool yesterday. I bought two different items. What items did I buy?

C. THE NEW SCHOOL CAFETERIA

1. Let's pretend our class is going to open a new restaurant in the school cafeteria. What kinds of foods would you like to see on the menu? (List them on the board.)

2. Let's put the prices we're going to charge next to each item. (The class can also decide on luncheon specials for each day of the week, discounts and special prices, and so on.)

3. On a piece of paper, design a menu for the new restaurant. (The class can decide on a name for the restaurant or each student can make up her own.)

D. FOOD FRACTIONS

1. (Draw a circle on the board. Say that it represents an apple— lemon, licorice, pizza, . . .—pie! Divide it into two, three, four, . . . equal parts and shade in one part.)

2. If I eat this much, what fraction of the pie have I eaten? (Shade another part.) Now what fraction have I eaten?

3. (Draw a rectangular casserole or cake. Divide and shade some parts and ask what fraction has been eaten.)

E. FOOD FOR THOUGHT

1. (Hold up empty grocery containers for everyone to see. They should have the weight or volume listed on the label.)

2. (Ask questions related to the containers and their contents:) How many grams do you think this holds? Which container do you think holds more? Weighs more?

F. TIPS

1. My friend and I ate at PittsBurgers Restaurant. The bill was $12.25 and we left a 10 percent tip—rounded to the nearest 5c. How much tip did we leave?

2. (Name other restaurant bills and have students figure 10 percent, 15 percent, and 20 percent tips—rounded to the nearest 5 cents, 10 cents, 25 cents, and so on.)

G. DIVIDING DESSERTS EVENLY

1. Please draw a rectangle (circle) to represent a pan of carrot cake (jello, granola crisps, pecan pie, . . .).

2. Divide the pan into 4 (6, 8, 12, . . .) equal parts.

H. I'LL TAKE HALF

1. (You'll need a recipe for this one. Use a cookbook, or ask students to bring their favorite recipe from home, or use the carob cake recipe:

Carob Cake Recipe (customary measure)	**Carob Cake Recipe** (metric measure)
¾ c whole wheat flour	180 mL whole wheat flour
1 tsp baking powder	5 mL baking powder
¼ tsp salt	1 mL salt
½ c butter or margarine	120 mL butter or margarine
½ c carob powder	120 mL carob powder
⅓ c honey	80 mL honey
2 eggs	2 eggs
⅔ c sunflower seeds	160 mL sunflower seeds
3 tbsp milk	45 mL milk
¼ tsp vanilla	1 mL vanilla
Bake at 350° for 15-20 minutes.	Bake at 350° for 15-20 minutes.

2. (List the ingredients on the board.)

3. How much of each ingredient will we need if we halve (double, . . .) this recipe?

I. WHAT'S THE ROASTING TIME?

1. (Write on the board:)

ROASTING CHART

	Minutes per Pound
Beef	
rare	20
medium	24
well	30
Chicken	15
Turkey	16
Lamb	35
Pork Loin	38
Veal	40

2. How long will it take to cook a five-pound turkey? Three-pound medium roast beef?

J. MENUS FROM MANY LANDS

1. Suppose we're all going to eat at an Italian (Mexican, Japanese, English, Kenyan, . . .) restaurant. Let's list on the board five items we would find on the menu.

2. What do you estimate each would cost? (List prices next to items.)

3. How much will it cost if *each* of us orders spaghetti? If I had ten quarters, could I buy the pepperoni pizza?

Traveling

Map-ematics Fun   Activity 7-7

C H A P T E R 7

We travel by car, bus, plane, train . . . subway, taxi, camel . . . wheelchair, roller skates, bicycle . . . and feet! Our lives are dependent on transportation, not only to get us where we're going, but also to get us things we need.

The activities in this chapter are your tickets to traveling with mathematics: How far is it? How shall we travel? Which route shall we take? What time must we leave? How much money will we need? In the following pages, you'll find activities related to each of these questions. They're presented as games, activity sheets, graphs, and even a vacation in your own neighborhood. If your students are feeling especially creative, have them make up their own travel games using the board and spinner on pages 280 and 282.

Everything you'll need for a safe and exciting trip is included in the pages of this chapter. According to our schedule, it's time to pack your bags, stretch your imagination, get your tickets, and enjoy these mathematical adventures. Bon voyage!

Let's Take a Conservation Vacation

Students calculate the expenses involved in a one-week holiday spent in their hometown.

Math Skills:	**calculating costs • addition • multiplication • estimation**
Curricular Areas:	**reading • social studies • art**
Materials:	**travel brochures for *your* town • newspapers • poster paper • crayons or markers**

Students dream about faraway places. But here's a chance for them to see the fun and excitement that is available in their own backyards.

PREPARATION: Ask the local chamber of commerce, automobile association, and real estate offices for literature about your town.

DISCUSSION: What kinds of things would tourists enjoy seeing and doing in our town? What expenses would you have on a trip to a faraway place that you don't have sightseeing in your own town? (hotel, taxi, gas)

DIRECTIONS:
1. Pretend you're a tourist who has just arrived in town. Here are some brochures and information about our town. Take 10 or 15 minutes to look them over and find things you'd like to do.

2. On the board, let's list the things you'd like to do and the places you'd like to go. (museum, movie, play, sports event, concert, park, library, open air market, shopping mall, or university)

3. Beside each, we'll put the entrance fee.

4. Each of you is going to plan a seven-day vacation around town. divide your paper into seven rows. Starting with Sunday, write one day of the week in each space. On the right-hand half of the paper, make four columns. Label them Entrance Fee, Food, Souvenirs, Other.

5. From the list on the board, choose at least one activity for each day of the week and fill it in on your chart.

6. In the appropriate columns, write what your expenses would be for each day's trip.

7. Draw a picture to show each day's activity.

8. Total the columns to find the cost of your vacation. (Have students check their answers by estimating.)

9. How have we saved money by taking this vacation here rather than a thousand miles away? How have we saved energy?

Trips to the Zoo

Students solve math problems related to gas consumption.

Math Skills:	subtraction • multiplication • comparison • measurement
Curricular Areas:	reading • social studies
Materials:	Activity Sheet 7-2.

Carpooling is a realistic real-world topic for problem solving. Students can make up their own problems or use this Activity Sheet for practice.

DIRECTIONS: (Have students complete Activity Sheet 7-2.)

(The answers to the problems are: 1. A. big car, moped, 13 gallons; 1. B. motorcycle; 1. C. compact car; 2. A. 8 gallons, 40 gallons; 2. B. 32 gallons, 160 gallons.)

Trips to the Zoo

1. Pat Wolfe is a volunteer at the zoo. Her family has several vehicles and she uses a different one each day. Here is her gas record for last week:

Monday	motorcycle	3 gallons
Tuesday	compact car	8 gallons
Wednesday	pickup	10 gallons
Thursday	big car	15 gallons
Friday	moped	2 gallons

A. Which vehicle used the most gas? _____

   The least? _____ What's the difference? _____

B. Which used more gas—the moped or the motorcycle? _____

C. Which used less gas—the pickup or the compact

   car? _____

2. Four of the zookeepers drive Cheetah compact cars and live in the same neighborhood. Each car uses 8 gallons of gas going to and from work.

A. When they carpool, how much gas do they use in a day?

   _____ In five days? _____

B. When they don't carpool, how much gas do

   they use in a day? _____

   In five days? _____

A Graphic Vacation

Students discuss places they'd like to go on vacation. They then choose their favorite places and use different kinds of graphs to record their choices.

Math Skills: making graphs

Curricular Areas: social studies • art

Prerequisites: bar graph • scattergram • circle graph • pictograph (see Chapter 1)

Materials: 1-cm graph paper (page 284)

This activity gives students a chance to try out their favorite graphs. And once it's done, you'll have some exciting student work to display on the most prominent bulletin board you can find.

PREPARATION: Plan to divide the class into small groups (three to five students) to work on the activity.

DISCUSSION: What are graphs used for? What kinds of graphs have we learned how to use? Why are there so many different kinds of graphs?

DIRECTIONS:

1. We're going to use different kinds of graphs to show our choices of vacation spots. What are some places that you'd really like to go for vacation? (Select four of the places that students name and list them on the board.)

2. How shall we determine which of these vacation places each person likes best? (As students show their preferences, someone records the numbers on the board.)

3. Now, we'll work in small groups. Each group will make a graph that shows the information on the board. (Assign one type of graph to each group.)

4. (Encourage the students to discuss and experiment with rough drafts before trying to complete their graphs. Some students may want to write computer programs for their graphs.)

5. (When the graphs are completed, have each group show its finished graph and explain the process they used to make it. They can also comment on whether or not the kind of graph they were assigned is appropriate for this information.)

Time for a Travel Game

Using information from a bus schedule, students play a game to see who arrives at a specified destination in the fewest moves.

Math Skills: calculating with time • addition • subtraction • making charts and tables

Curricular Areas: reading

Materials: copies of CARE-VAN Bus Timetable (page 175) • game spinner (page 280), paper clip and pencil • paper for tally sheet • (optional: local bus, train, or plane schedules)

Bus or train schedules inspire an unusual math game.

PREPARATION:

1. Make one copy of the game spinner (page 280). Divide each section in half. Duplicate one copy for each group of players.

2. Each group fills in the ten sections on the spinner with different time intervals, such as 10 min, 5 h, 1 h 30 min, 50 min, 2 h, 30 min, 1 h, 3 h 45 min, 15 min, 7 h 30 minutes.

DISCUSSION:

Suppose we leave St. Louis on Monday to go to Phoenix. What time does the bus leave St. Louis? (4:15 a.m.) Where and when is the next stop (Tulsa, 1:05 p.m.) Is it still Monday? (Yes) Where and when is the next stop? (Amarillo, 11:25 p.m.) Is it still Monday? (Yes) Next stop? (Albuquerque, 4:50 a.m.) What day is it now? (Tuesday)

DIRECTIONS:

(two to four players)

1. The players in each group decide what trip they will take. Choose a starting point and a destination. Read the schedule and write the starting point and departure time at the top of your tally sheets (Pittsburgh, 12:20 p.m.). At the bottom, write your destination and the scheduled arrival time (St. Louis, 3 a.m.).

2. First player spins. He records the time shown on the spinner (1h 30 min) on his tally sheet and adds it to his departure time to find the new time (1:50 p.m.).

3. Play continues to the left. For each play, the player spins and adds that amount to the time on his tally sheet.

4. First person to reach or pass the arrival time for the destination is the winner.

VARIATIONS: Use current bus, train, or plane schedules for cities in your area.

Bus Timetable (Use for Activity 7-4.)

CARE-VAN BUS TIMETABLE
We CARE-ry You Across the Continent

Day 1	(ET)	Quebec	leave	12:05 p.m.
	(ET)	Boston	arrive	7:15 p.m.
			leave	8:00 p.m.
Day 2	(ET)	New York	arrive	12:40 a.m.
			leave	1:45 a.m.
	(ET)	Philadelphia	arrive	3:50 a.m.
			leave	4:05 a.m.
	(ET)	Pittsburgh	arrive	11:35 a.m.
			leave	12:20 p.m.
	(ET)	Cincinnati	arrive	7:15 p.m.
			leave	7:45 p.m.
	(ET)	Indianapolis	arrive	9:35 p.m.
			leave	10:20 p.m.
Day 3	(CT)	St. Louis	arrive	3:00 a.m.
			leave	4:15 a.m.
	(CT)	Tulsa	arrive	1:05 p.m.
			leave	2:15 p.m.
	(CT)	Amarillo	arrive	11:25 p.m.
			leave	11:55 p.m.
Day 4	(MT)	Albuquerque	arrive	4:50 a.m.
			leave	5:30 a.m.
	(MT)	Phoenix	arrive	2:50 p.m.
			leave	3:45 p.m.
Day 5	(PT)	Las Vegas	arrive	12:05 a.m.
			leave	12:50 a.m.
	(PT)	Los Angeles	arrive	8:20 a.m.
			leave	9:15 a.m.
	(PT)	San Diego	arrive	11:40 a.m.
			leave	12:35 p.m.
	(PT)	Tijuana	arrive	1:20 p.m.

ET = Eastern Time   CT = Central Time
MT = Mountain Time   PT = Pacific Time

Charge on Down the Road

Students use a set of gas credit slips to calculate distance and gas consumption for a trip.

Math Skills:	ordering numbers • addition • subtraction • multiplication • division • measurement calculations • calculating costs
Curricular Areas:	reading
Materials:	Activity Sheet 7-5 • gas credit slips (page 180) • (optional: calculators • scissors)

This activity provides a tankful of math practice—as well as a close look at the expense involved in traveling by car!

PREPARATION: Before duplicating the Activity Sheet, fill in the current price of a liter of gas at the top of the page.

DISCUSSION: Use problems such as the following to review the problem-solving steps needed for this activity.

1. Cost of gas: If the price of gas is $_____ per liter, and you buy 42.5 liters, how much will you have to pay?

2. Kilometers traveled: If your odometer said 25802 at the start of a trip and 25962 at the end, how far did you travel? (160 km)

3. Kilometers per liter: If you used 34.8 liters of gas to travel 435 kilometers, how many kilometers per liter did you get? (12.5)

DIRECTIONS: 1. (Have students complete Activity Sheet 7-5.)

2. (Optional: Have students cut apart the credit slips and put them in order before doing step 1 on the Activity Sheet.)

VARIATION: Have students make up computer programs that could be used at the gasoline pump to calculate the cost of a certain quantity of gas at a given price.

ChargeOn Down the Road

The price of unleaded gas is $_____ per liter.

Materials needed: Gas Credit Slips

1. Number the six gas credit slips in order from the first purchase to the last. The dates and odometer readings will help you.

2. On each credit slip, fill in the price of gas. Then calculate the total amount for each purchase.

3. On the chart below, record the dates, odometer readings, liters of gas, and purchase prices.

4. Calculate and record the kilometers traveled between each gas purchase.

5. Calculate and record the kilometers per liter for each gas purchase.

6. Calculate the totals for kilometers traveled, gas purchased, and total price of gas. Calculate the average kilometers per liter for this period of time.

GAS RECORD SHEET September 8–25					
DATE	ODOMETER READING	KILOMETERS TRAVELED	LITERS OF GAS PURCHASED	KILOMETERS PER LITER	TOTAL PURCHASE PRICE
TOTALS				AVERAGE:	

Now that you're an expert, you could volunteer to keep track of the distances and gas expenses for the family car!

Slip 1 (top left)

623 53749 195
LEE GARCIA
Ling Ping Gas
4321 Main St.
Guzzletown 09090

Premium ☐ Regular ☐ Unleaded ☐

Amount

SUPER CITY GAS

Qty	Price	Amount

Date _____
X _____
Customer Signature

Odometer Reading _____

Slip 2 (top right)

623 53749 195
LEE GARCIA
Ling Ping Gas
4321 Main St.
Guzzletown 09090

Premium ☐ Regular ☐ Unleaded ☐

Amount

SUPER CITY GAS

Qty	Price	Amount

Date _____
X _____
Customer Signature

Odometer Reading _____

Slip 3 (bottom left)

623 53749 195
LEE GARCIA
Ling Ping Gas
4321 Main St.
Guzzletown 09090

Premium ☐ Regular ☐ Unleaded ☐

Amount

SUPER CITY GAS

Qty	Price	Amount

Date _____
X _____
Customer Signature

Odometer Reading _____

Slip 4 (bottom right)

623 53749 195
LEE GARCIA
Ling Ping Gas
4321 Main St.
Guzzletown 09090

Premium ☐ Regular ☐ Unleaded ☐

Amount

SUPER CITY GAS

Qty	Price	Amount

Date _____
X _____
Customer Signature

Odometer Reading _____

Bus, Train, or Plane?

Students calculate and compare the cost of using different forms of transportation for a family trip.

Math Skills: calculating costs • addition • multiplication • working with decimals, fractions, and percentages

Curricular Areas: reading • social studies

Materials: Activity Sheet 7-6 • (optional: calculators)

Calculating costs before you leave home is one of the best ways to make sure a trip doesn't fail because of unexpected expenses.

PREPARATION: Copy the fares below on the board. (Please note that adult fares are given in dollar amounts, children's fares in fractions, and senior citizens' in percentages. You may want to choose which arithmetic operations you want to emphasize.)

FARES	YOUR TOWN TO THE EMERALD CITY		FARES
	Bus	Train	Plane
Adult	$25.83	$36.50	$52.00
Child (2–12)	½ adult fare	Alone: ½ adult fare With parent: ¼ adult fare	$42.00
Under 2	Free	Free	Free
Senior Citizen	13% off adult fare	25% off adult fare	$52.00

DISCUSSION: If your whole family takes a trip, do you take the bus, a train, a plane, or go by car? How do you decide which means of transportation to use? (cost, time, comfort, convenience, and so on)

DIRECTIONS:
1. Before you do your Activity Sheet, you need to decide how many people are going on this trip. Plan the trip for at least four people.

2. (Have students complete Activity Sheet 7-6.)

VARIATIONS:
1. Have students call a travel agent to get actual prices for a trip from your town to an interesting vacation spot. They can use the chart on the Activity Sheet to calculate the cost by bus, train, and plane. To calculate the cost by car, they can figure out the distance, then use a current mileage rate to estimate.

2. Have students write computer programs to calculate special fares (child, senior citizen) based on regular adult fares.

Bus, Train, or Plane?

You and your family are planning a trip from your town to the Emerald City. Just for fun, include some grandparents, cousins, or friends. You have volunteered to compare the costs of traveling by bus, by train, and by plane.

Use the fares your teacher gives you. Fill in the charts below and calculate how much it would cost your family to travel in three different ways.

HOW MANY PEOPLE?

_____ Adults   _____ Children (2 to 12)

_____ Senior Citizens   _____ Children (under 2)

		How Many?	Price per Ticket	Total Price
B	Adults			
U	Children (2 to 12)			
S	Senior Citizens			
			Total Cost	

T				
R	Adults			
A	Children (2 to 12)			
I	Senior Citizens			
N			Total Cost	

_____

_____

P | _____

L Adults

A Children (2 to 12)

N Senior Citizens

E   Total Cost

Find the difference between the total cost for bus and train. _____

Train and plane. _____ Plane and bus. _____

Map-ematics Fun

Students do a variety of activities using maps of their hometown, state or country. (*See photo, page 167.*)

Math Skills: These six activities involve one or more of the following skills: addition • subtraction • estimation • linear measurement • scale measurement • making charts and tables • working with coordinate geometry and spatial relations

Curricular Areas: reading • social studies

Materials: maps (one for each group of three or four students) • (optional: calculators)

Put a map in the hands of your students and you've got their attention. And don't worry! Maps are easy to get from your chamber of commerce, a real estate office—or see page 268. Of course, you can always get just one map, and then make copies for everyone.

PREPARATION: If there are not enough maps to go around, make copies to distribute to the class.

(A giant wall map can be made with the help of an overhead projector. Make a transparency of the map you'd like to enlarge. Hang a large piece of paper and trace around the transparency image projected by the overhead. Students can use the map on the wall. This is even more fun if you make the map on a sheet and then put it on the floor for students to walk on.)

DISCUSSION: In what ways have you used maps? Why are maps necessary?

DIRECTIONS: Give your students 10 or 15 minutes to look at their maps before starting any of these activities. As you do an activity, try to include the places on the map that they seem particularly interested in.

RUNNING IN A MAP-ATHON

Skills: addition

Materials: maps that show mileage between towns or other points

1. (Pick two places on the map.) Suppose you are running in a map-a-thon to raise money for your community center. Your route goes between _____ and _____ . Find the total distance between these two points.

2. Here's an extra challenge. Where could the map-a-thon be if it was exactly 26 miles? (Give various distances depending on the map.)

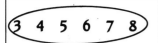

GEO-MAP-TRY FUN

Skills: coordinate geometry

Materials: maps that have letters and numbers along the edges

1. Find the airport (or other landmark) on your map. Using the letters and numbers along the edge of the map, can you describe its location? (Name other places such as an intersection of two streets, a school, or a town, and ask for the coordinates of that location. Also ask for two pairs of coordinates between which runs a particular river, highway, railroad, and so on.)

2. What park is at C,6? What street, starting with the letter *E,* is at G, 2? (Name other coordinates and a clue of the kind of place found there.)

GEO-MAP-TRY GAME

Skills: coordinate geometry; addition; subtraction

Materials: maps that have letters and numbers along the edges

1. (Several games can be going on in your room at once. Two students or teams of two to three students can play each other.)

2. Will each person or team please make a list of five questions and answers giving both coordinates and places. Then make another copy of your questions (for your opponent to use).

3. At the signal, switch questions with the opposing player or team. Race to find the answers.

MAP-PROXIMATE THE DISTANCE

Skills: scale measurement; estimating distances

Materials: maps with distance scales

1. (Pick two places on the map.) Use the scale to find how far it is between _____ and _____ . (Repeat this exercise several times, using different places on the map.)

2. (Name two new places on the map.) Estimate the distance between _____ and _____ . Then check to see how close you came.

WHAT'S YOUR MILEAGE MAP-TITUDE?

Skills: charts and tables

Materials: maps with a mileage chart

1. (Pick two places.) What is the distance between _____ and _____ ? (Repeat this exercise several times, using different places on the mileage chart.)

Chapter 7: Traveling

2. What is the shortest distance shown on the chart? The longest?

A MAP-BELIEVE TOWN

Skills: coordinate geometry; scale measurement; spatial relations; measurement—linear

Materials: drawing paper and markers

1. Design a map of a make-believe town you'd like to live in. The town could be based on a book, television program, or movie. Include streets, rivers, parks, and other landmarks.

2. Mark off numbers and letters along the edge of the map. Then make an index indicating the coordinates for different places on the map.

3. Make a scale of the map and indicate mileage between points.

Quick and Easy

A. CATEGORIES FOR COMMUTING

1. (Have students suggest 20 forms of transportation. List them on the board. For instance, skateboard, camel, airplane, ship, and so on.)

2. How can we group these kinds of transportation into categories? (land, sea, and air; size; power source; number of wheels; and so on)

3. (Have students regroup them into different categories.)

B. HOW FAST DOES SUPERMAN FLY?

1. (Name a distance—in miles or kilometers—and an amount of time—in hours. Have the students calculate the miles per hour—MPH—or the kilometers per hour—KPH.)

2. (For fun, make up a story surrounding this information. Television, movie, or book characters can be traveling to places in their stories.)

C. MPG and KPL R E-Z 2 Cal-Q-Late!

1. (Miles per gallon—MPG—and kilometers per liter—KPL—are easy to calculate. Just name an amount of gas and a distance.) I traveled 216 miles and used 9 gallons of gas. Calculate the MPG for my car.

2. (Have students round their answers to the nearest whole number—tenth, and so on.)

D. EXOTIC FIELD TRIPS

1. Suppose we were taking a special field trip to see the animals at world-famous Mighty Moose Animal Resort. Our bus is scheduled to leave here at 7:30 a.m. and arrive there at 11 a.m. How long will it take us to get there?

2. (Suggest other exotic field trips, such as:)

	Depart	Arrive
Daredevil Dome Entertainment Park	11:15 a.m.	2:45 p.m.
Scrumptious Strawberry Farms	5:17 a.m.	6:08 a.m.
Gladstone's Gold Mine	8:05 p.m.	10:20 p.m.
Mystery Mountain	3:48 p.m.	5:12 p.m.

E. PARALLEL AND PERPENDICULAR STREETS

1. Who can name a street that is parallel to the street our school is on? What's another parallel street?

2. What streets are perpendicular to our street?

3. Can anyone name a street that's parallel to the street you live on? One that's perpendicular?

F. GETTING TO THE ROUTE OF THINGS

1. (Put four—or five, six, . . .—points on the board and connect them in some way. Use at least five lines. Have the students copy this drawing.)

2. Suppose these lines represent a highway system. Start at any point. Can you find a route that will take you over each highway only once?

3. (Have a student draw another similar puzzle on the board. Ask the students to predict whether or not they'll be able to trace the entire highway system on this puzzle.)

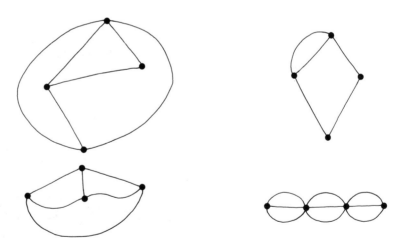

Part 3.
Projects

Projects

A Consumer Report   Project 8-4

1. *A Class Party*

2. *A First Class Supermarket*

3. *Behind Supermarket Scenes:*
   A Field Trip

4. *A Consumer Report:*
   Paper Towels

5. *Classroom Post Office*

6. *A Catalog Wishing Center*

Real-world math covers many issues, many topics, and many techniques for problem solving. In this chapter, a different approach to problem solving is presented in six comprehensive projects. Although the emphasis is on math, these projects are perfect to be easily integrated with other curricular areas such as social studies, science, and language arts.

These projects take time and effort to set up. And they require a week or more to carry out. But when you introduce one of these projects in your class, you'll be teaching skills and concepts from several areas of the curriculum—all at the same time. And you'll be giving your students a unique opportunity to experience the real world in a very special way.

A Class Party

Students will increase a recipe to make enough for the entire class, then calculate the cost of their ingredients. If you like, they can also prepare the food for a class party.

Math Skills: working with fractions • metric and customary measurement— volume • addition • multiplication • calculating costs

Curricular Areas: reading • handwriting • vocabulary • science • home economics

Materials: Activity Sheet 8-1 • recipes (page 198) • (optional: mixing bowls, spoons, metric and customary measuring equipment, blender, oven)

We've offered delicious recipes and actual directions for a class party. However, this activity is also fine if you just want to do the calculations.

PREPARATION:
1. Make one copy of the recipe page (page 198) and four copies of Activity Sheet 8-1. Cut out the recipes and fasten one to each Activity Sheet.

2. Divide the class into four groups.

3. Provide a list of prices for all the ingredients, or have students find the prices in the newspaper or at the supermarket.

DISCUSSION: How does a chef (or our cafeteria cook) know how much food to prepare for a large crowd? How can we figure out how much to prepare when we cook for the whole class?

DIRECTIONS: (Calculations only)
1. (Have students complete Activity Sheet 8-1.)

(An actual party)
1. (Assign each recipe to a group of students.)

2. (Provide the necessary ingredients—a small group might help shop.)

3. (Assign each group a time to prepare their food.)

4. Let's eat!

VARIATIONS:
1. Choose just one recipe and write it on the board. Distribute Activity Sheet 8-1 to the class. Have all students increase the recipe and calculate the cost of the ingredients.

2. Have students make up computer programs to increase recipes to make enough for the whole class. If a printer is available, students will enjoy having a hard copy of their class-size recipes.

A Class Party

You will need a copy of a recipe for this activity.

Name of recipe _____

The recipe is for _____ people or servings.

Our class has _____ students.

We'll have to increase the recipe _____ times to have enough for everyone in the class.

On the lines below, list the ingredients in the recipe. Next to each ingredient, write the quantity needed. Then fill in the number of boxes, pounds, cans, and so on you will need to buy and the price of each. Calculate the cost for all the ingredients.

Recipe for _____

INGREDIENT	QUANTITY	HOW MANY BOXES, POUNDS, CANS, AND SO ON	PRICE OF EACH	TOTAL COST

GRAND TOTAL: _____

Recipes
(Use for Activity 8-1).

Gorp

2 C. raisins
½ C. sunflower seeds
½ C. walnuts
½ C. coconut
2½ C. peanuts

Put all the foods into a bowl. Toss them until mixed. Serves 12.

Fruit Shake

½ banana
½ C. frozen strawberries
½ C. milk

Blend in blender for 15 seconds on high.
Makes one fruit shake.

Deviled Eggs

6 hard boiled eggs
1 tbsp. mustard
3/4 C. mayonnaise
2 tbsp. pickle vinegar

Peel eggs. Cut them in half. Add yolks to above ingredients. Mix well. Scoop mixture into empty "bowls" of the eggs.
Makes 12 servings

Cheese Cucumber Sandwich

2 slices rye bread
2 tsp. mayonnaise
¼ C. cottage cheese
4 slices of cucumber
4 radish slices

Toast the bread. Spread mayonnaise on each slice. Spread cottage cheese on one slice. Place the cucumber and radish slices on top of the cottage cheese. Cover with the other slice of toast. Makes 1 sandwich.

A First-Class Supermarket

Students create a first-class supermarket using empty supermarket containers and play money.

Math Skills: numbers and numeration • addition • subtraction • multiplication • division • classifying • counting money • making change • estimation • geometry

Curricular Areas: reading • vocabulary • social studies • science • art • home economics

Materials: bottles, boxes, cans, and sacks from the supermarket • play money • (optional: poster paper • crayons or markers • register tapes • magazines and newspapers • calculators)

Once this is set up in your room, it'll be a constant reminder to students (as well as parents and other visitors!) that math is really a real-world activity.

PREPARATION: Ask students to bring empty containers from the supermarket (see sample letter to parents—page 257). Identify an area of the room that can be used for the class supermarket. Appoint some students to "invent" a cash register.

There are many ways to create and use a classroom supermarket. These suggestions are presented in a weekly format so that you can introduce new activities and skills selectively. As the weeks go by, the operation of the supermarket becomes more and more similar to that of a real-world supermarket. By this time, your students will *really* be ready for Project 3 (see page 202).

You may want to devote one or two math classes each week to supermarket activities. Or you may want to rotate small groups so that each student works in the supermarket at least twice a week.

And for teachers who love a project and don't mind a change of pace, just plunge right in and let the supermarket *be* your math class!

DISCUSSION: If you were a supermarket manager, how would you decide what products to carry? Decide what prices to charge? Arrange your store? Attract customers?

DISCUSSION: If you were a supermarket manager, how would you decide what products to carry? Decide what prices to charge? Arrange your store? Attract customers?

DIRECTIONS: WEEK 1: Collecting

Ask students to bring supermarket containers from home. Assign committees to decorate the supermarket area, select a name for the store, put tables and shelves in place, and so on.

WEEK 2: Sorting and Classifying

Ask a group of students to arrange the items in the supermarket. Have them explain the rationale for their categories. On the following days, have other students arrange the items as they wish and explain their classifications. By the end of the week, have the class agree on a good arrangement.

WEEK 3: Money Practice

In preparation for opening the supermarket, practice counting money to make change.

WEEK 4: Number Practice

Advertise a grand opening for the supermarket. Have students design posters to display around the room. Each poster should include some numerical information (dates, hours, quantities, prices).

WEEK 5: Making Change

Open the supermarket! Have each student purchase just one item. Students take turns at the register so that everyone gets to practice making change.

WEEK 6: Finding Totals

Students purchase three or four items. Each student gets a turn at the register for practice in adding and making change.

WEEK 7: Numbers on Packages

Have students examine the way items are packaged. Which packages really serve a useful purpose? What numerical information is given on the package?

WEEK 8: Estimation

Students purchase at least six items. By rounding off to the nearest ten, they estimate the price of their purchases before going to the checkout counter. The cashier finds the exact total and compares it with the customer's estimate.

WEEK 9: Prices and Discounts

Students make a flyer or newspaper ad. Use old magazines and newspapers for pictures of items needed in the ads. Include special offers (coupons) for next week's sales.

WEEK 10: Calculator Calculations

Students make purchases using the special offers from last week's ads. Cashiers use calculators at the checkout counter.

WEEK 11: Geometry

Have students look at the shapes of the boxes, bottles, and cans in the supermarket. Why are some round? Rectangular? Tall? Flat? Have each student design a new-style container for some common supermarket item.

WEEK 12: Computerized Register Tapes

(Save your computerized register tapes for this one.) Distribute register tapes to the students. Discuss the information they show. On a sheet of paper, students write directions such as:

> Draw a line under the date.
> Draw a ring around the total.
> Draw a red line through a taxable item.
> Write the brand name of the tuna.
> List three produce items.

Students staple their register tapes to their directions and exchange with each other to do the activities.

WEEK 13: A Lucky Week!

By this time, you're probably way ahead of us! While the supermarket is still in respectable shape, why not have a visitors' day, or invite a reporter from the local paper to interview the class and take pictures. All that work—and all that fun—can lead to some good public relations for your school.

Behind Supermarket Scenes: A Field Trip

Students help plan a field trip to a supermarket. Then, at the supermarket, they follow instructions on Task Cards. After the field trip, they follow up on their activities.

Math Skills:	calculating costs, weight, and time • addition • comparison • geometry
Curricular Areas:	reading • writing • vocabulary • social studies • art • home economics
Materials:	Task Cards 1–13

One carefully planned trip to the supermarket provides a delightful experience that teaches real-world math in a way that will be long remembered.

PREPARATION: Call the manager of a supermarket to arrange a field trip. It will help if you're prepared to explain the purpose of the trip and the kinds of things students will be looking for (see the Task Cards). Ask the manager to greet the students and to plan a short talk (three to five minutes) to acquaint them with the store.

DIRECTIONS: Activities at School:

1. (Do one or two of the supermarket activities described in Chapter 5, Shopping.)

2. (Discuss the characteristics of supermarkets. Have students draw or write about their impressions of a supermarket.)

3. (Have students list things they want to find out at the supermarket. To make the questions more useful on the field trip, help students group their questions and ideas in some logical manner—by sections of the supermarket, by persons responsible, and so on.)

4. (Have students make drawings to show how they think the supermarket is, or should be, laid out. Discuss the reasons for their store arrangements. Have each student write down two or three questions about the store layout that they will try to answer during the field trip.)

Task Cards for the Field Trip:

1. (Divide the students into groups of two or three. Assign one or two Task Cards to each group. Duplicate the number of Task Cards needed for your groups.)

2. (Task Cards 1 through 10 can be done by several different groups.)

3. (Task Cards 11, 12, and 13 are interviews. If you plan to use these, arrange for the interviews with the appropriate people.)

Follow-Up Activities:

1. (Use questions such as the following to help students review what they learned at the supermarket:)

   A. What was the most interesting thing you saw or learned at the supermarket?

   B. In what ways was the supermarket like other supermarkets you have seen? In what ways was it different?

   C. If you were going to work at the supermarket, what job would you like to have? Why?

   D. What are the ways numbers are used at the supermarket?

   E. What kinds of math do people who work at the supermarket have to know?

2. (Have students write thank-you letters to the manager and other people they met at the supermarket. Encourage students to describe one special thing they learned or enjoyed on the trip.)

3. (If several students did the same Task Card, have them compare the information they collected. Did they all write down the same information? Why or why not?)

4. (Students can work in small groups to make posters, drawings, or three-dimensional displays to illustrate something they observed or learned at the supermarket. This would be a good way to share the information they recorded on their Task Cards.)

5. (Have students who did Task Cards 1, 2, 3, and 4 make a chart or graph to show the price information they collected. Have them present the information to the class and discuss questions such as:)

   A. Are store brand names the least expensive?

   B. Are the most expensive items always the well-known brands?

   C. Are the most expensive items the highest quality?

   D. Are the most expensive items the most attractively packaged?

6. (Have students who interviewed an employee at the supermarket dramatize what that person does. After they act out the role, they can tell about additional information they found out such as hours and salary.)

Copyright © 1986 by Addison-Wesley Publishing Company, Inc.

Least Expensive Brands

(Use with Project 8-3.)

You are a very careful shopper with a limited budget. Look for the least expensive brand or kind of the following items. Write the brand name and price of each item on the lines:

BRAND NAME   PRICE

eggs—1 dozen

milk—1 quart or
       1 liter

margarine—1 pound
       (450 grams)

peanut butter—8 ounces
       (225 grams)

bread—1 loaf

tomato soup—10¾ oz
       (305 grams)

frozen peas—20 oz
       (567 grams)

cold cereal—1 pound
       (450 grams)

facial tissues—200

liquid detergent—32 oz
       (946 mL)

TOTAL $ _____

Most Expensive Brands

(Use with Project 8-3.)

You are a shopper who thinks that the most expensive item is always the best buy. Find the most expensive brand of the following items. Write the name and price of each on the lines.

BRAND NAME   PRICE

eggs—1 dozen   _____

milk—1 quart or
        1 liter   _____

margarine—1 pound
        (450 grams)   _____

peanut butter—8 ounces
        (225 grams)   _____

bread—1 loaf   _____

tomato soup—10¾ oz
        (305 grams)   _____

frozen peas—20 oz
        (567 grams)   _____

cold cereal—1 pound
        (450 grams)   _____

facial tissues—200   _____

liquid detergent—32 oz
        (946 mL)   _____

                                    TOTAL $ _____

Shopping in a Hurry

(Use with Project 8-3.)

You are in a big hurry, so you buy the first kind of each item that you see. Find the following items, take a quick look, and write the brand name and price of the ones you see first.

	BRAND NAME	PRICE

eggs—1 dozen

milk—1 quart or
      1 liter

margarine—1 pound
      (450 grams)

peanut butter—8 ounces
      (225 grams)

bread—1 loaf

tomato soup—10¾ oz
      (305 grams)

frozen peas—20 oz
      (567 grams)

cold cereal—1 pound
      (450 grams)

facial tissues—200

liquid detergent—32 oz
      (946 mL)

TOTAL $ _____

Packaging

(Use with Project 8-3.)

You really like products that are wrapped in attractive packages.
For each of the following items, find the kind that looks best.
Write the brand name and price on the lines below.

BRAND NAME   PRICE

eggs—1 dozen _____

milk—1 quart or
   1 liter _____

margarine—1 pound
   (450 grams) _____

peanut butter—8 ounces
   (225 grams) _____

bread—1 loaf _____

tomato soup—10¾ oz
   (305 grams) _____

frozen peas—20 oz
   (567 grams) _____

cold cereal—1 pound
   (450 grams) _____

facial tissues—200 _____

liquid detergent—32 oz
   (946 mL) _____

TOTAL $ _____

Prices of your Favorite Brands

(Use with Project 8-3.)

Before you go to the supermarket, write down your favorite brands of the items on the list below. At the supermarket, compare the prices of these brands with other brands. Check the column to show if your brand is the most expensive, the least expensive, or the in-between price of that item.

	ITEM	FAVORITE BRAND	MOST EXPENSIVE	MIDDLE-PRICED	LEAST EXPENSIVE
1.	Cereal	_____			
2.	Crackers	_____			
3.	Juice	_____			
4.	Bread	_____			
5.	Soup	_____			
6.	Chips	_____			
7.	Hot dogs	_____			
8.	Catsup	_____			
9.	Peanut butter	_____			
10.	Ice cream	_____			

Colors of your Favorite Brands

(Use with Project 8-3.)

Before you go to the supermarket, write down your favorite brands of the items on the list below. Find your brands at the supermarket and check the colors of their labels. Write the two most noticeable colors for each on the lines.

	ITEM	(AT SCHOOL) FAVORITE BRAND NAME	(AT THE SUPERMARKET) MOST NOTICEABLE	
			COLOR 1	COLOR 2
1.	Cereal	_____	_____	
2.	Juice	_____	_____	
3.	Bread	_____	_____	
4.	Peanut butter	_____	_____	
5.	Soup	_____	_____	
6.	Hot dogs	_____	_____	
7.	Catsup	_____	_____	
8.	Chips	_____	_____	
9.	Ice cream	_____	_____	

Which three colors are used most often?

1. _____

2. _____

3. _____

Why do you think these colors are used so often? _____

_____

_____

Designing Labels

(Use with Project 8-3.)

Pretend that you are an artist who designs labels for food products. Your task is to examine designs and colors used on labels. Look at the labels on ten different products. What color is used for the background? What color is used most in the foreground?

List the products you examine on the lines; put the number of each in the correct square on the scattergram:

1. _____

2. _____

3. _____

4. _____

5. _____

6. _____

7. _____

8. _____

9. _____

10. _____

Foreground Color						
Other						
Yellow						
Green						
Brown						
Red						
White						
	White	Red	Brown	Green	Yellow	Other

Background Color

Which color is used most often for the background? _____

Which color is used most often for the foreground? _____

Which combination of colors is used most often? _____

Shapes in the Supermarket

(Use with Project 8-3.)

You are a geometry expert. Find the following shapes in the store. Try to find at least two items for each shape. Write the names of the items next to the shapes.

1. Rectangle   _____   _____

2. Circle   _____   _____

3. Cylinder   _____   _____

4. Triangle   _____   _____

5. Square   _____   _____

6. Trapezoid   _____   _____

7. Sphere   _____   _____

8. Rectangular solid   _____   _____

| TASK CARD | 9 | Name: _____ |

Get to Know your Store

(Use with Project 8-3.)

Find the items listed below in the supermarket. Beside each one write the number or letter of the aisle where you found it. Write the brand name, size, and price for one of each of the items.

ITEM	AISLE	BRAND NAME	SIZE	PRICE
1. Mustard	_____	_____		
2. Canned peaches	_____	_____		
3. Salt	_____	_____		
4. Spaghetti	_____	_____		
5. Cottage cheese	_____	_____		
6. Paper towels	_____	_____		
7. Toothpaste	_____	_____		
8. Orange juice	_____	_____		

The Electronic Scanner

(Use with Project 8-3.)

You shop at a store that uses an electronic scanner. You are very curious about the accuracy of the scanner. Ask a cashier for a register tape that a customer left on the checkout counter. Go through the store to find items listed on the tape and write their names below. Beside each, write the price shown on the tape, the price shown on the item, and the price shown on the shelf. (You may want to focus on items that are <u>on sale</u> this week, as there may be more chance for error here.)

ITEM	PRICE ON REGISTER TAPE	PRICE ON ITEM	PRICE ON SHELF
1. _____	_____	_____	_____
2. _____	_____	_____	_____
3. _____	_____	_____	_____
4. _____	_____	_____	_____
5. _____	_____	_____	_____
6. _____	_____	_____	_____
7. _____	_____	_____	_____
8. _____	_____	_____	_____
9. _____	_____	_____	_____
10. _____	_____	_____	_____

Interview—Cashier

(Use with Project 8-3.)

Interview a cashier. Ask the questions below. Make up another question of your own.

1. What are your duties? _____

   _____

   _____

   _____

2. How do you work the cash register? _____

   _____

   _____

   (You may draw a picture on the back to show the register.)

3. What are your working hours? _____

4. How many hours do you work each day? _____

5. How many days do you work each week? _____

6. What is the beginning salary for a cashier? _____

7. (Your own question:) _____

   _____

   _____

   _____

Interview—Butcher

(Use with Project 8-3.)

Interview a butcher. Ask the questions below. Make up another question of your own.

1. What does a butcher do? _____

   _____

   _____

   _____

2. What equipment is needed? _____

   _____

   _____

3. What are your working hours? _____

4. How many hours do you work each day? _____

5. How many days do you work each week? _____

6. What is the beginning salary for a butcher? _____

7. (Your own question:) _____

   _____

   _____

   _____

Interview—Produce Worker

(Use with Project 8-3.)

Interview someone who works in the produce section. Ask the questions below. Ask more questions if you wish.

1. What does a produce worker do? _____

   _____

   _____

   _____

2. How do you keep the fruits and vegetables fresh? _____

   _____

   _____

   _____

3. What are your working hours? _____

4. How many hours do you work each day? _____

5. How many days do you work each week? _____

6. What is the beginning salary for someone in the produce

   section? _____

7. (Your own question:) _____

   _____

   _____

   _____

A Consumer Report: Paper Towels

Students conduct quality tests on four brands of paper towels. They record and compile data, then discuss their results in relation to strength, durability, and absorbency of paper towels. (*See photo, page 193.*)

Math Skills: collecting, recording, and interpreting data • metric measurement—mass and volume • addition • comparison

Curricular Areas: reading • social studies • science • home economics

Materials: four different brands of paper towels (include those familiar brown school towels!) • scouring powder • window cleaner • ten transparent 100-mL containers (with 10-mL markings) • metric weights of 5 g, 10 g, 50 g, 100 g, 500 g (at least five of each) (The high school science teacher may be able to lend you containers and/or weights) • water • trash cans or plastic bags for used paper towels • Data Record Forms 1–5 and Data Summary Form (pages 221 to 226)

Quality testing is often shown in commercials to convince us of the superiority of a particular product. This activity describes one way to compare the quality of different paper towels. You can also use it as a model for your class to test other products such as soap, toys, or cereals.

PREPARATION: When you purchase the paper towels, choose a different design or color for each brand. For each brand, tear off one towel, staple it to the plastic wrapper, and label it with a code letter: A, B, C, and D. (Keep these hidden in your desk so you'll know which towel is which brand when the test is completed!)

For each of the five activities, each student will need one towel of each brand. (Multiply five times the number of students to find how many towels of each brand are needed.)

Set up the activities in five stations so that four to six students can work at each station at one time. Volunteer helpers (parents, student assistants) can be very helpful in setting up and supervising the learning stations.

Each station should be equipped with:

1. A paper towel of each brand for each child

2. Data Record Forms

3. Trash can or plastic bag for used towels

Additional Material for Each Station:

STATION 1: STRENGTH

Metric weights—at least five each: 5 g, 10 g, 50 g, 100 g, 500 g (if more weights are available, students won't have to wait so long for a turn)

STATION 2: DURABILITY

Scouring powder

Water

STATION 3: DURABILITY

Window cleaner

STATION 4: ABSORBENCY

Five 100-mL containers

Water

STATION 5: ABSORBENCY

Five 100-mL containers

Water

DISCUSSION: What do you use paper towels for? Which paper towels are best for doing these things? When you help with the shopping, how do you decide which brand of paper towels to buy? (our favorite, the cheapest, the one we have a coupon for, and so on) What are some of the qualities a good paper towel should have? (strength, durability, and absorbency)

DIRECTIONS: Introducing the Stations:

1. We're going to conduct quality tests on four brands of paper towels. We'll do a different test at each of the stations. Each of you should do each test, then record your results on your own Data Record Forms. When we've finished all the stations, we'll compile this data and discuss what we've found out.

2. (Assign four to six students to each station—plus a volunteer if possible. Each group will need about twenty minutes at each station.)

3. At each station, you will find copies of the Data Record Forms. Please follow the directions and record your data carefully.

Summarizing the Data:

1. Let's look at the data we've collected. Will each of you please complete your Data Summary Form? (With younger students, you may want to do this a step at a time with the entire class.)

2. Now we need to add all our scores for all the towels. On this chart (or the chalkboard), I have labeled a column for each towel. Will the students in the first group please tell me their scores. (Record the scores as the students say them, or have them write their scores on the chart. When all scores are recorded, have the class total the scores for each towel.)

3. Which towel has the highest score? Does that mean it's the best towel? Why or why not? Which has the lowest score? Would you ever buy this towel? (Help students understand that many factors go into such decisions and that the results of one test are not necessarily the deciding factor.)

4. And now (. . . drum roll . . .) which brand is Towel A???

Station 1: Strength

(Use with Project 8-4.)

You will test the strength of each paper towel by placing weights in the center of a towel and gently bouncing them up and down until the towel tears.

1. Take one paper towel.

2. Place one or two small weights in the center.

3. Hold the towel on each side; bounce the weights up and down gently.

4. One at a time, add weights until the towel tears.

Rate the STRENGTH of each towel:

1. Very weak

2. Weak

3. Okay

4. Strong

5. Very strong

STRENGTH

TOWEL A	TOWEL B	TOWEL C	TOWEL D

Station 2: Durability

(Use with Project 8-4.)

You will test the durability of each kind of towel by using a towel with scouring powder to clean a desk or table top.

1. Shake a small amount of scouring powder on the top of the desk (or table).
2. Dampen the paper towel slightly.
3. Use the towel to clean the desk.
4. Examine the towel and decide how well it survived this task.

Rate the DURABILITY of each towel:

1. Completely disintegrated
2. Fell apart a little
3. Okay
4. Durable
5. Very durable

DURABILITY—With scouring powder

TOWEL A	TOWEL B	TOWEL C	TOWEL D

Chapter 8: Projects

Station 3: Durability

(Use with Project 8-4.)

You will test the durability of each kind of towel by using it to clean a window.

1. Spray a small amount of window cleaner on a window.
2. Use a paper towel to clean the window.
3. Examine the towel and decide how durable it is.

Rate the DURABILITY of each towel:

1. Completely disintegrated
2. Fell apart a little
3. Okay
4. Durable
5. Very durable

DURABILITY—Cleaning windows

TOWEL A	TOWEL B	TOWEL C	TOWEL D

Station 4: Absorbency

(Use with Project 8-4.)

This test is to determine how much water a paper towel will absorb.

1. Put 100 mL of water into one of the marked containers.
2. Hold a towel in the center and dip it in the water until it's completely soaked. Carefully pull it out and let it drip over the container.
3. How much water is left in the container?

   How much water did the towel absorb?

Rate the ABSORBENCY of each towel:

1. Not absorbent
2. Slightly absorbent
3. Okay
4. Did a pretty good job
5. Soaked up lots of water!

ABSORBENCY—100-mL container

TOWEL A	TOWEL B	TOWEL C	TOWEL D

Station 5: Absorbency

(Use with Project 8-4.)

At this station, you will determine how well a towel absorbs spilled water.

1. Pour 20 mL of water onto the table top (or other assigned place).
2. Wipe it up with your towel. Try to get the area completely dry.
3. By observing the area and the paper towel, decide how absorbent the towel is.

Rate the ABSORBENCY of each towel:

1. Left lots of water
2. Only slightly absorbent
3. Okay
4. Good towel for spills
5. Super absorbent

ABSORBENCY—Wiping up a spill

TOWEL A	TOWEL B	TOWEL C	TOWEL D

Data Summary Form

(Use with Project 8-4.)

	TOWEL A	TOWEL B	TOWEL C	TOWEL D
1. STRENGTH				
2. DURABILITY (scouring powder)				
3. DURABILITY (cleaning windows)				
4. ABSORBENCY (100-mLcontainer)				
5. ABSORBENCY (wiping up a spill)				
TOTAL				

Classroom Post Office

Task Cards describe math activities that occur in the daily operation of the post office. These, along with activities for the whole class, provide real-life experiences related to communicating by mail.

Math Skills: addition • subtraction • multiplication • numbers and numeration • ordering numbers • reading charts and tables • comparison • making change

Curricular Areas: reading • writing • social studies • art

Materials: Task Cards 1–6 • postage stamps (page 64) • postage rates charts (page 233–or obtain from your local post office) • index cards

---

Post office activities offer a great opportunity for sending "good work" notes to all your students.

PREPARATION:

1. Select the Task Cards you want to use and prepare the materials described on the cards.

2. (For Task Card 1) Make index card address labels using a variety of addresses that include several on the same streets. One address should be written on each card. Some suggested addresses are:

402 Window Street	1132 Back Wall Alley	64 Carpet Corner
405 Window Street	1139 Back Wall Alley	93 Carpet Corner
408 Window Street	1146 Back Wall Alley	324 Carpet Corner
411 Window Street	1153 Back Wall Alley	576 Carpet Corner
414 Window Street	1160 Back Wall Alley	845 Carpet Corner
417 Window Street	1167 Back Wall Alley	
4751 Teacher Terrace	7635 First Row Avenue	8300 Second Row
4759 Teacher Terrace	7645 First Row Avenue	8315 Second Row
4767 Teacher Terrace	7655 First Row Avenue	8330 Second Row
4774 Teacher Terrace		8345 Second Row
		8360 Second Row
17843 Chalkboard Circle		
17906 Chalkboard Circle		

DIRECTIONS:

1. HOLIDAY GREETINGS: Numbers and numeration

   (Any holiday is a great occasion to use the real post office.) We are going to design greeting cards or write letters to relatives or friends. When your card or letter is finished, we'll address an envelope and mail it. Please bring the address from home. Why is it important to write the address very neatly? What are the different ways numbers are used in the addresses?

2. SECRET FRIENDS: Numbers and numeration

I have put all your names—and mine—in this box. We'll draw names. The name you draw is your secret friend for a week. Each day, please write a thoughtful note to your secret friend. It will be delivered through the class post office. On the last day of the week, we'll sign our names so we can find out who has been our secret friend.

3. STAMP-OUT GAME: Geometric shapes; spatial relations

(Students play a game where they learn to use a variety of problem-solving strategies. See page 61.)

4. TASK CARD ACTIVITIES

(You can use any one of these six Task Cards as a lesson for the whole class. Or you can set them up as activities for a post office learning station.)

House Numbers

(Use with Project 8-5.)

MATH SKILLS: Numbers and numeration; ordering numbers

MATERIALS: Index cards with addresses

Addresses can be assigned to students. They can make address signs to display on their desks for use in delivering class mail.

DIRECTIONS: Sort the address cards by street. For each street, put the addresses in numerical order. Then sort each pile into odd numbers and even numbers.

Zip Codes

(Use with Project 8-5.)

MATH SKILLS: Ordering numbers

MATERIALS: Index cards (ten or more) with a different zip code on each one.

DIRECTIONS: Sort the cards by putting them in order from lowest to highest zip code number.

First-Class Postage

(Use with Project 8-5.)

MATH SKILLS: Addition; multiplication; reading charts

MATERIALS: First-class postage rate chart

DIRECTIONS:

Find the first-class postage for the following items:

1. One 3-ounce letter

2. One regular letter; three single postcards

3. Four letters, each less than 1 ounce

4. Seven single postcards

5. One 2-ounce letter; one 3-ounce letter; and three 1-ounce letters

Make up more first-class postage problems. Write them neatly on a paper so that other students can solve them.

Parcel Post Rates

(Use with Project 8-5.)

MATH SKILLS: Addition; multiplication; reading charts

MATERIALS: Parcel post rate chart

DIRECTIONS:

Find how much postage is needed to send the following:

1. One package—3 lbs 2 oz—to Zone 3

2. One package—less than 2 lbs—to Zone 8

3. Two packages—each 4 lbs—to Zone 5

4. Two packages—each 37 lbs 6 oz—one to Zone 2, one to Zone 6

5. One package—12 lbs—to Zone 4

6. Two packages—each 6 lbs—to Zone 4

Look at problems 5 and 6. Which costs more—sending one large package or two smaller packages?

Using the rate chart, make up your own parcel post problems to share with the class.

Parcel Post Rates—
Different Zones

(Use with Project 8-5.)

MATH SKILLS: Subtraction; reading charts; comparison

MATERIALS: Parcel post rate chart

DIRECTIONS: Write the answers to these questions on a separate piece of paper.

1. Look at the column for LOCAL rates.

   What is the difference between the rates for:

   a. 2 lbs and 3 lbs?

   b. 3 lbs and 4 lbs?

   c. 4 lbs and 5 lbs?

   d. 5 lbs and 6 lbs?

   e. 6 lbs and 7 lbs?

   f. 7 lbs and 8 lbs?

2. Look at the column for ZONE 5 rates.

   What is the difference between rates for:

   a. 20 lbs and 21 lbs?

   b. 21 lbs and 22 lbs?

   c. 22 lbs and 23 lbs?

   d. 23 lbs and 24 lbs?

   e. 24 lbs and 25 lbs?

   f. 25 lbs and 26 lbs?

Parcel Post Rates— Different Weights

(Use with Project 8-5.)

MATH SKILLS: Subtraction; reading charts; comparison

MATERIALS: Parcel post rate chart

DIRECTIONS: Write the answers to these questions on a separate piece of paper.

1. Look at the rates for a package weighing between 9 and 10 pounds.

   What is the difference between the rates for:

   a. Local and Zones 1 and 2?

   b. Zones 1 and 2 and Zone 3?

   c. Zone 3 and 4?

   d. Zone 4 and 5?

   e. Zone 5 and 6?

   f. Zone 6 and 7?

   g. Zone 7 and 8?

2. Look at the rates for a package weighing between 41 and 42 pounds.

   What is the difference between the rates for:

   a. Local and Zones 1 and 2?

   b. Zones 1 and 2 and Zone 3?

   c. Zone 3 and 4?

   d. Zone 4 and 5?

   e. Zone 5 and 6?

   f. Zone 6 and 7?

   g. Zone 7 and 8?

Postage Rates Charts (Use for Project 8-5.)

EXPRESS MAIL — NEXT DAY SERVICE
FIRST CLASS

LETTER RATES:

1st ounce.......................................22¢
Each additional ounce.........................17¢

For Pieces Not Exceeding (oz.)	The Rate Is	For Pieces Not Exceeding (oz.)	The Rate Is
1	$0.22	7	$1.24
2	0.39	8	1.41
3	0.56	9	1.58
4	0.73	10	1.75
5	0.90	11	1.92
6	1.07	12	2.09

FOR PIECES OVER 12 OUNCES SEE FIRST-CLASS ZONE RATED (PRIORITY) MAIL RATES

CARD RATES:

Single Postal Cards sold by the post office	14¢ each.
Double postal cards sold by the post office	28¢ (14¢ each half.)
Single post cards	14¢ each.
Double post cards (reply-half of double post card does not have to bear postage when originally mailed)	28¢ (14¢ each half.)
Presort rate	Consult Postmaster
Business reply mail	Consult Postmaster

FIRST-CLASS ZONE RATED (PRIORITY) MAIL

Weight over 12 ounces and not exceeding-pound(s)	Local zones 1, 2, and 3	Zone 4	Zone 5	Zone 6	Zone 7	Zone 8
1	$ 2.40	$ 2.40	$ 2.40	$ 2.40	$ 2.40	$ 2.40
2	2.40	2.40	2.40	2.40	2.40	2.40
3	2.74	3.16	3.45	3.74	3.96	4.32
4	3.18	3.75	4.13	4.53	4.92	5.33
5	3.61	4.32	4.86	5.27	5.81	6.37
6	4.15	5.08	5.71	6.31	6.91	7.66
7	4.58	5.66	6.39	7.09	7.80	8.67
8	5.00	6.23	7.07	7.87	8.68	9.68
9	5.43	6.81	7.76	8.66	9.57	10.69
10	5.85	7.39	8.44	9.44	10.45	11.70
11	6.27	7.97	9.12	10.22	11.33	12.71
12	6.70	8.55	9.81	11.01	12.22	13.72
13	7.12	9.12	10.49	11.79	13.10	14.73
14	7.55	9.70	11.17	12.57	13.99	15.74
15	7.97	10.28	11.86	13.36	14.87	16.75
16	8.39	10.86	12.54	14.14	15.75	17.75
17	8.82	11.44	13.22	14.92	16.64	18.76
18	9.24	12.01	13.90	15.70	17.52	19.77
19	9.67	12.59	14.59	16.49	18.41	20.78
20	10.09	13.17	15.27	17.27	19.29	21.79
21	10.51	13.75	15.95	18.05	20.17	22.80
22	10.94	14.33	16.64	18.84	21.06	23.81
23	11.36	14.90	17.32	19.62	21.94	24.82
24	11.79	15.48	18.00	20.40	22.83	25.83
25	12.21	16.06	18.69	21.19	23.71	26.84
26	12.63	16.64	19.37	21.97	24.59	27.84
27	13.06	17.22	20.05	22.75	25.48	28.85
28	13.48	17.79	20.73	23.53	26.36	29.86
29	13.91	18.37	21.42	24.32	27.25	30.87
30	14.33	18.95	22.10	25.10	28.13	31.88
31	14.75	19.53	22.78	25.88	29.01	32.89
32	15.18	20.11	23.47	26.57	29.90	33.90
33	15.60	20.68	24.15	27.45	30.78	34.91
34	16.03	21.26	24.83	28.23	31.67	35.92
35	16.45	21.84	25.52	29.02	32.55	36.93
36	16.87	22.42	26.20	29.80	33.43	37.93
37	17.30	23.00	26.88	30.58	34.32	38.94
38	17.72	23.57	27.56	31.36	35.20	39.95
39	18.15	24.15	28.25	32.15	36.09	40.96
40	18.57	24.73	28.93	32.93	36.97	41.97

FOURTH CLASS

(PARCEL POST) ZONE RATES
CONSULT POSTMASTER FOR WEIGHT AND SIZE LIMITS

NONMACHINABLE SURCHARGE:
A parcel mailed to a ZIP Code destination outside the BMC service area for your post office is subject to a surcharge of $.090 in addition to the rate shown in this table if:
- A. It is nonmachinable according to the standards prescribed in Domestic Mail Manual section 753 or
- B. It weighs more than 35 pounds.

WITHIN (INTRA-BMC) BMC DISCOUNT:
A parcel mailed to a ZIP Code destination shown below is for delivery within the BMC service area for your post office and is eligible for a discount of $0.16 from the rate shown in this table.

WITHIN (INTRA-BMC) BMC ZIP CODE DESTINATION FOR YOUR POST OFFICE ARE:

Weight 1 Pound and not exceeding (pounds)	Local	Zones 1-2	Zone 3	Zone 4	Zone 5	Zone 6	Zone 7	Zone 8
2	1.35	1.41	1.51	1.66	1.89	2.13	2.25	2.30
3	1.41	1.49	1.65	1.87	2.21	2.58	2.99	3.87
4	1.47	1.57	1.78	2.08	2.54	3.03	3.57	4.74
5	1.52	1.65	1.92	2.29	2.86	3.47	4.16	5.62
6	1.58	1.74	2.05	2.50	3.18	3.92	4.74	6.49
7	1.63	1.82	2.19	2.71	3.51	4.37	5.32	7.36
8	1.69	1.90	2.32	2.92	3.83	4.82	5.91	8.25
9	1.75	1.99	2.46	3.13	4.15	5.26	6.49	9.12
10	1.80	2.07	2.59	3.34	4.48	5.71	7.07	10.00
11	1.85	2.13	2.70	3.49	4.71	6.03	7.49	10.62
12	1.90	2.20	2.79	3.65	4.94	6.34	7.89	11.23
13	1.94	2.26	2.89	3.79	5.15	6.63	8.27	11.78
14	1.98	2.32	2.98	3.92	5.35	6.90	8.62	12.30
15	2.02	2.37	3.06	4.04	5.54	7.15	8.94	12.79
16	2.06	2.42	3.14	4.16	5.71	7.39	9.25	13.24
17	2.10	2.48	3.22	4.27	5.88	7.61	9.54	13.67
18	2.14	2.52	3.29	4.38	6.03	7.83	9.81	14.08
19	2.18	2.57	3.36	4.48	6.18	8.03	10.07	14.46
20	2.21	2.62	3.43	4.58	6.33	8.22	10.32	14.83
21	2.25	2.67	3.49	4.67	6.47	8.41	10.56	15.18
22	2.28	2.71	3.56	4.76	6.60	8.59	10.79	15.51
23	2.32	2.75	3.62	4.85	6.73	8.76	11.01	15.83
24	2.35	2.80	3.68	4.94	6.85	8.92	11.22	16.14
25	2.39	2.84	3.74	5.02	6.97	9.08	11.42	16.43
26	2.42	2.88	3.80	5.10	7.08	9.23	11.61	16.72
27	2.45	2.92	3.85	5.18	7.20	9.38	11.80	16.99
28	2.48	2.96	3.91	5.25	7.30	9.52	11.98	17.26
29	2.52	3.00	3.96	5.33	7.41	9.66	12.16	17.52
30	2.55	3.04	4.01	5.40	7.51	9.80	12.33	17.77
31	2.58	3.08	4.07	5.47	7.61	9.93	12.50	18.01
32	2.61	3.12	4.12	5.54	7.71	10.06	12.66	18.24
33	2.64	3.15	4.17	5.61	7.81	10.19	12.82	18.47
34	2.67	3.19	4.22	5.68	7.90	10.31	12.97	18.70
35	2.70	3.23	4.26	5.74	7.99	10.43	13.12	18.91
36	2.73	3.26	4.31	5.81	8.08	10.54	13.27	19.12
37	2.76	3.30	4.36	5.87	8.17	10.66	13.41	19.33
38	2.79	3.33	4.41	5.93	8.25	10.77	13.55	19.53
39	2.82	3.37	4.45	5.99	8.34	10.88	13.69	19.73
40	2.85	3.40	4.50	6.05	8.42	10.98	13.83	19.92

United States Government Charts

A Catalog Wishing Center

Six small activity centers are set up in the room as "wishing centers." Students follow directions on Task Cards that ask them to order items that will add pizzazz to the room from catalogs. They order furniture, decorations, food and drink items, games and activities, music and TV items, and instructional equipment.

Math Skills: addition • subtraction • multiplication • division • measurement

Curricular Areas: reading • spelling • handwriting • vocabulary • social studies • art • home economics

Materials: Task Cards A–F • a supply of catalogs • order forms (page 42 or actual ones from the catalogs)

Let the students' imaginations run wild with this activity. They'll love putting their daydreams into real terms—they'll even forget they're doing math!

PREPARATION:

1. Identify six areas where you'll want the activity centers. They can be six student desks—or one long table with six chairs.

2. Fill in an appropriate budget amount on each Task Card.

3. Set a Task Card at each place. Next to each put one or two appropriate catalogs and a supply of order forms.

DISCUSSION:

What would you do if you had permission to redecorate our classroom? What things would you buy?

DIRECTIONS:

1. Follow the instructions on the Task Card at the center.

2. You will be asked to fill out a catalog order form. Be sure to:
   - Stay within the given budget;
   - Order at least five items;
   - Include catalog numbers, prices, and totals on the form.

3. (It might work well to have each student complete a different center each week, or . . .)

4. (If you have enough catalogs, duplicate the Task Cards and make enough centers for the whole class at once!)

Task Card A

DECORATION COMMITTEE

Use a catalog and order form to buy a variety of items to decorate the room.

Order items for the walls and floors, and even items for the top of each student's desk!

Measure carefully to be sure the items will fit where you want them.

Order at least five items.

Task Card B

FOOD AND DRINK COMMITTEE

Use a catalog and order form to buy appliances and equipment you would need to cook and serve food in the class (microwave, store, refrigerator, sink unit, pots, utensils, glasses, and so on.

Measure carefully to be sure the items will fit where you want them.

Order at least five items.

FURNITURE COMMITTEE

Determine how many student desks you will keep.

Use a catalog and order form to buy additional furniture for the room including couches, chairs, tables, and bookcases or wall shelving.

Measure carefully to be sure the items will fit where you want them.

Order at least five items.

Task Card D

GAMES AND ACTIVITIES COMMITTEE

Use a catalog and order form to buy fun activities for the room. Include such things as pool and table tennis equipment, aquariums, books, video games, and puzzles.

Measure carefully to be sure items will fit where you want them.

Order at least five items.

Task Card E

INSTRUCTIONAL EQUIPMENT COMMITTEE

Use a catalog and order form to buy equipment that can be used in the classroom to teach math. Include such items as a movie or slide projector, projection screen, typewriter, mimeograph machine, calculators, and computers.

You may order individual items for every student, as long as you stay within your budget.

Order at least five items.

Task Card F

MUSIC AND TV COMMITTEE

Use a catalog and order form to buy a television, radio, stereo system, tape deck, records, tapes, guitar, drums, and so on.

You can also order the necessary tables and shelves on which to put the equipment.

Measure carefully to be sure the items will fit where you want them.

Order at least five items.

Part 4.
Suggestions for
Using This Book

Tips for Teachers

Go Fly a Kite! Activity 3-7

This book is designed primarily to be used as a supplement to your math textbook. In addition, because of its focus on a broad range of real-world topics, it will be very useful as a supplementary resource in other areas of the curriculum, especially social studies, science, reading, and language arts. It can also be a valuable source of ideas for integrated units of study such as conservation, consumer education, or career awareness.

The content and scope of this book make it adaptable to many uses. The following section outlines some of those uses, offers ideas on how to get started, and gets specific about choosing (and changing) an activity to suit your own classroom needs.

Using the Activities

Ease of use has been an important goal in writing this book. The Table of Contents and the quick-reference tables following it are worth getting to know—they will be your best sources of ideas to guide you to appropriate activities.

When it comes to ease of use, we think one of the best features of the book is the concise format we've used for all activities. Once you get acquainted with this format, you'll be able to glance at any activity and find just the information you want.

Although each activity is self-explanatory, we do have some general ideas about using the activities that we'd like to share with you.

Choosing an Activity. As we indicated earlier, we've grouped the activities in chapters by skill areas and real-world topics. Thus, they are not arranged in a sequential order from the front of the book to the back. We encourage you to select them at random, jumping around the book to find activities that suit your needs.

One good way to choose an activity is to start by finding a topic that relates to other class activities—or one that simply appeals to you. Let's assume that your fifth-grade class has been learning about stores and shops in the neighborhood and that you're emphasizing addition practice. A quick glance at the Table of Contents suggests that a good topic would be Shopping (Chapter 5). As you turn to this chapter, you look for activities that include addition at the fifth grade level (shown in the grade level indicator at the top of the page).

When you thumb through the chapter, you'll see that Activities 1, 2, 3, 7, 8, 9, and 10 all include addition practice. By reading the brief descriptions of these activities, you can select one (or more!) that suits your needs.

Many activities can be adapted to cover a wide span of achievement, and thus are indicated for many grade levels. For instance, Activity 5-2, Cash Register Tape Arithmetic, is recommended for grades 3 through 8. If this is the activity you choose, you can select register tapes that involve addition at the level your class can handle.

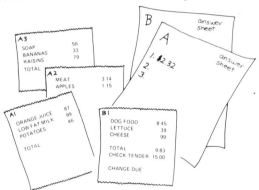

Once you find an activity that fits the topic and skills you need, you can make the level of difficulty just right for your students: change the numbers, vary the difficulty of discussion questions, omit some of the steps, or include one of the variations.

At times, you'll find yourself choosing an activity that includes a skill your students haven't yet learned. Don't be afraid to go ahead and use it. For example, if your students haven't learned to add decimals, you might still use a problem where they add money ($1.65 + $2.20). And even if they haven't learned division, you can be sure they won't have any trouble dividing 27 peanuts among 3 friends!

One of the wonderful things about real-world activities is that sometimes students get so caught up in them they simply forget they don't know how to do the math. They just go ahead and try different ways to solve the problem—certainly a valuable approach for dealing with real-world problems!

Some ways you can help your students to deal with unfamiliar skills are:
- Use real-world objects to count
- Use a calculator
- Use the process of trial and error

Planning the Activity. Once you've chosen an activity, the next step is to see what's suggested for MATERIALS and PREPARATION. Most of the printed materials that you'll need are included right with the activity. Page numbers are always indicated so the materials are easy to find.

Many of the real-world materials included in the book require prices on them. On most of these, we've left blank spaces where the prices go. This way, you can choose numbers to make the problems more or less difficult to suit the achievement levels of your students. For instance, if you want to use amounts less than a dollar, fill in the blanks with cents only. If you want to use easy numbers, you can fill in even dollar amounts. If your students need extra practice in regrouping, you can use numbers that require regrouping.

An added bonus from filling in your own prices is that you can always have current, realistic math lessons. Regardless of what happens to the price of gas, or even dog food, your real-world materials will always be up to date!

The MATERIALS section lists other supplies in addition to the printed items needed for the activity. You'll be able to find most of these items in your classroom supply closet.

Dear Parents:

Our class will soon be studying _____ in math class. In order to do some real-world activities about this topic, we will be using some extra materials.

We would really appreciate your help. Would you please save your _____ and send them to school with your child? We need them by _____ .

Thanks for your help.

Sincerely,

_____

(Use school stationery if possible)

Dear _____ :

Our class will soon be studying _____ in math class. In order to do some real-world activities about this topic, we will need some extra materials.

Please help us. Would you donate _____ ?
We would be glad to pick them up at your convenience. We need them by _____ .

I will be calling you on _____ to answer any questions you may have. If you would like to talk to me before that, please call me at _____ .

Thank you for your help.

Sincerely,

_____

Sometimes, we've suggested additional materials such as newspapers, empty supermarket containers, and travel brochures that are easy to obtain from local sources. Chapter 10 includes suggestions for collecting such materials.

Another step in planning an activity is deciding how to group your students. Most of the activities are designed to be used with the whole class, but they're equally useful for smaller groups. Other activities such as games are designed for small groups. In most of these cases, it's a simple matter to divide the whole class into small groups—and let them all play the game at the same time.

A few activities consist of Task Cards to be used in learning stations. There are also some activities that work best for independent study. These can be made available at an interest table or as part of a learning station.

As you can see, there's a lot of flexibility in the ways you can use these activities. We encourage you to let your plans reflect your own style so you'll enjoy these real-world activities as much as your students do.

Adapting Activities. If your class is typical, your students will have a wide range of math abilities. One of the strengths of the activities in this book is that they can easily be adapted to different levels of difficulty. There are several ways to do this:

- Choose only part of an activity
- Vary the difficulty of computations
- Have students use calculators
- Emphasize different skills
- Vary the level of questions about a problem

For example, in a single activity you could ask four different questions, depending on your students' abilities:

- Which number is greater?
- How much greater is this number than that one?
- What fraction of one number is the other?
- What percent of one is the other?

Another feature of these activities is that many of them can be used more than once. There are several good reasons for doing this:

- For repeated practice in the same skill
- For increasingly difficult practice in a skill
- To apply a skill in a different context

When you repeat an activity, it doesn't have to be exactly the same each time. You can give it a new twist by:

- Changing the numbers
- Including one of the VARIATIONS

- Changing the topic or context
- Using a different set of real-world materials (such as menus obtained from a local restaurant)

Once you find activities that are particularly appealing to your students, you may find yourself doing them over and over. And why not? Your students will be enjoying themselves, and they'll be learning math, too!

Using Activities in Related Curricular Areas. Although the main focus of this book is math, most of the activities can be used in other curricular areas. For instance, as a social studies lesson you might want to use Don't Get Caught in a Graphic Jam (Chapter 1); a good activity for science would be A Watched Pot Always Boils (Chapter 2); or The Shape of the News (Chapter 3) could be used for an art lesson.

An easy way for you to cover more bases during your busy day is to combine the time allotted for two different curricular areas. Then you can choose a real-world activity that fits both these areas, such as Map-ematics Fun (Chapter 7). Students will then experience how skills needed in the real world require different kinds of learning.

To make it easier for you to find appropriate activities, we've included a matrix on page xi that shows curricular areas related to each activity.

Special Features

We've included a number of special features to make the book as flexible and useful as possible.

Quick and Easy Activities. Do you sometimes have a few extra minutes before dismissal time? Do you find yourself looking for something to do while waiting for the last bus to arrive? Every activity chapter has Quick and Easy activities that are designed for such times.

We suggest that you look them over and mark several that are especially appealing to you. Then, when those free minutes occur, you'll be ready with a catchy activity that's just right for your class.

You'll soon discover that many of the Quick and Easy activities can be used over and over. For some of them, you can repeat the same questions using different numbers. For others, you can repeat the same process, but use different content. For example, after you've used the process of building a bar graph to show your students' favorite cars, you can use the same process to show favorite pets, snacks, colors of shoes, numbers of pencils in desks—and on and on and on!

Real-World Math Games. Because games are such a good way to practice and learn math, we've included at least one in each chapter. The list on the inside back cover will be a handy reference when you and your class are looking for a game.

Real-World Materials. Using real-world materials is an important and exciting part of real-world math. You may want to plan activities around a certain real-world material. You can scan the Real-World Materials Chart on the inside front cover to get a general idea of inexpensive and easy-to-obtain materials you can use. Then dip into the specific suggestions for Materials and Resources in Chapter 10 for an abundance of ideas about how you can enrich your math classes with a specific real-world material and related guest speakers and field trips.

Problem-Solving Approach. We've made a special effort to emphasize realistic products and problems. When your students do these activities, they will have to take a problem-solving approach, by considering:

- What am I trying to find out?
- What information do I have?
- Is it enough? Too much?
- How can I get additional information?
- How do I use this information to solve the problem?
- Is my answer reasonable?
- How can I check it?

In general, the activities are much more comprehensive than typical story problems. This makes it easy to emphasize different aspects of problem solving such as estimation, choosing which calculation skill to use, and identifying reasonable solutions when there is no single right answer.

Use of Computers. The use of computers in the classroom continues to undergo rapid change. Because of this we have included computers as useful, but not essential, tools of instruction for real-world math.

We have included suggestions related to computer literacy, programming, and computer-assisted instruction.

Computer literacy is emphasized as a variation in activities about places that depend on computers for their day-to-day operations (such as banks and supermarkets).

For teachers who have enthusiastic young programmers in the classroom, we've provided suggestions for programming activities such as calculating average daily attendance in your school or using LOGO to create tessellations. There's really no limit to the programming possibilities that you and your students can create related to activities throughout the book.

In the area of computer-assisted instruction, a wealth of software related to real-world math continues to appear on the market. We have included a few specific suggestions related to available programs. But, for the most part, we leave this area up to you—and simply encourage you to integrate the software that you own with the activities in this book as much as possible.

Use of Calculators. Because calculators are so closely related to real-world math, they're ideally suited for use with real-world math activities in your classroom. We've indicated the best activities for this purpose by listing calculators in the MATERIALS section of those activities.

The decision of whether or not your students should use calculators really depends on the purpose of your lesson. When your purpose is to emphasize calculation practice, you'll probably choose not to have calculators available. Otherwise, you'll find that calculators are very helpful tools when you want to:

- Have students check their own calculations
- Adapt an activity for use at a lower grade level
- Emphasize the problem-solving process rather than calculation skills
- Be sure that all students are successful in some math activities

Measurement—Metric and Customary. The activities in Chapter 2 are all designed to help your students learn or improve their measurement skills. Most of these activities are written to be used with both customary and metric measurement, so you can choose the system that fits your curriculum. When you specifically want to emphasize the metric system, we suggest you try Activity 2 (A Metric Schoolroom), Activity 6 (Grocery Grams Galore), or Activity 9 (The Metric Marketplace Game).

Activities to Do at Home. As we indicated in the introduction to the book, one of the underlying themes of this book is promoting good relations between home and school. We encourage you to take advantage of opportunities to convey positive messages while you're teaching real-world math skills.

For some activities, students collect data at home: dimensions of a rug or the weight of a box of cereal. For other activities, students may be asked to bring newspapers, magazines, or empty cracker boxes to use in math class.
Such interactions will help parents and other adults understand that homework can be more than just practice problems. They'll also get a better idea of what's going on at school—and they'll be delighted to see that their children are learning the real-world skills that will help them become wise consumers.

Special Uses

In the previous pages, we have shared many ideas for teachers who are using this book in their regular classrooms. In this section, we'll outline some other uses.

Staff Development. We think the book can be especially useful to curriculum specialists, resource teachers, principals, and others who have responsibility for continuing professional development in their schools. In a workshop setting, real-world math activities can be used very effectively to achieve several important goals:

- To provide teachers with examples of effective activities
- To give teachers a chance to experience activities before introducing them in their classes
- To provide the kind of information and support teachers need in their ongoing efforts to improve instructional programs and procedures

Substitute Teachers. We hope that substitute teachers will find this book especially useful. More and more, substitutes are carrying their own bag of plans with them—to avoid the situations that can occur when regular lesson plans aren't available or don't work. Real-world math activities can be particularly effective because they:

- Are highly motivating to students
- Can be adapted to different levels of difficulty
- Provide opportunities to practice as well as apply math skills

You can provide yourself with a set of flexible lesson plans by putting together the following materials:

- Thirty copies of menus (Activity 6-1)
- Thirty copies of order blanks (Activity 6-1)
- Two sets of game boards (page 282)—one set blank, one set filled in
- A set of 15 maps or 30 copies of a single map (Activity 7-7)
- Thirty sheets of graph paper (page 284 or 285)
- A collection of register tapes (Activity 5-2)

With these few basic materials and some well-thought-out lesson plans, you'll be all set for that 6 a.m. phone call at any time of the year!

Part 5.
Materials
and Resources

Materials and Resources

Specific Suggestions for Materials and Resources (Newspapers, page 268; Supermarket Items, page 274.)

Menus, ads, credit slips, labels. Every day we deal with an abundance of real-world materials. And because real-world math is what this book is all about, it too deals with an abundance of materials.

To save you time, we've included many authentic-looking items throughout the book. (See the chart on the inside front cover.) There are menus, a catalog page and order form, bus timetables, and more. But you can add even more realism by using items from places right in your hometown, such as order forms from a local restaurant, pages from the catalog of a local store, and bus time tables from your own city bus line.

Included in this chapter are some hints about ways to get materials (from local sources), ways to use materials (your materials as well as ours), guest speakers, and field trips. We've even thrown in a suggestion for using telephone technology as an inexpensive alternative to a guest speaker or a field trip!

We hope these materials are just the magic ingredient you need to help your students enjoy math while they experience and solve realistic real-world problems.

General Suggestions for Materials and Resources

There are a variety of ways for you to obtain and use materials and resources to bring real-world math into your classroom This section contains a few general suggestions just to get you started.

Ways to Get Materials. The following are tried and true ways to obtain materials with a minimum of trouble and expense.

- Ask parents to help by saving and donating materials. The sample letter on page 257 may help you.
- Request items, in writing, from local businesses. And don't forget to *use school stationery*. It works every time! In fact, your school secretary may be willing to type your letters for you. If that secretary is busy, ask the superintendent's secretary! (Don't laugh . . . just go ask!) The sample letter on page 258 may help you.
- Many times the easiest way is to collect materials here and there a little at a time. When you have enough of a certain item, which may be next month or next year (!), you'll be ready to do the activity.

- If you can only get one copy of a particular item, duplicate it to make a copy for each student.
- A very important thought: *Sharing materials with your fellow teachers saves time!* If you collect menus . . . and another teacher gets some maps . . . and then you share . . . you've each cut your time in half! Of course, if you get a third teacher to get some train schedules . . . (!)

SAMPLE LETTER TO PARENTS

Dear Parents:

Our class will soon be studying _____ in math class. In order to do some real-world activities about this topic, we will be using some extra materials.

We would really appreciate your help. Would you please save your _____ and send them to school with your child? We need them by _____ .

Thanks for your help.

Sincerely,

_____

SAMPLE LETTER TO BUSINESSES

(Use school stationery if possible)

Dear _____ :

Our class will soon be studying _____ in
math class. In order to do some real-world activities about this
topic, we will need some extra materials.

Please help us. Would you donate _____ ?
We would be glad to pick them up at your convenience. We
need them by _____ .

I will be calling you on _____ to answer any
questions you may have. If you would like to talk to me before
that, please call me at _____ .

Thank you for your help.

Sincerely,

_____

Ways to Use Materials. See the Real-World Materials Chart on page
261 and the inside front cover. This chart lists materials that are in
this book and the pages where they're found. You can often substitute items from local firms for the materials we've given you in
the book.

If you laminate or cover your materials with clear contact paper,
they'll last much longer. Usually, this is a good investment of your
time and energy—however, this covering sometimes takes away from
the authentic look of the materials.

Many materials can be made into transparencies and used with an
overhead projector. Ads, maps, newspapers, price lists, and other
such materials can be used very effectively on an overhead. You can
ask questions or encourage students to make up questions.

Guest Speakers. Having someone come to your room to share information, experiences, and dreams can spark days of anticipation and excitement in your class. Here are some thoughts that we hope will be of help to you.

- Parents and grandparents are excellent resources as guest speakers.

- Many companies have public relations personnel who would be delighted to come to your school. It's good public relations for the business—and a great experience for your class.

- When you and your class decide on a guest speaker, have your students decide what questions they want to ask. Then when you call to make arrangements with the speaker, you can share this list of questions.

- Questions might revolve around the speaker's career—why she chose it, if she likes it, what her career goals are, and so on.

- Questions can also focus on working hours, salaries, duties and responsibilities, schooling required, and interesting experiences on the job. (You might want to caution the class not to ask questions that are too personal. "What is the beginning salary for your career?" is a less personal question than "How much do you make?")

Field Trips. Do you remember going on field trips when you were young? We do. And they stand out as important events in our school experiences. Even though it may only have been a class walk to the corner store, a restaurant, or a bank, the lasting impression made by that learning experience is seldom duplicated by any other school experience. Here are some ideas to keep in mind as you plan a field trip.

- Local businesses and industries are excellent places to visit. Sometimes they're as close as a phone call and a leisurely walk! If you telephone a fast-food restaurant, a supermarket, or even a bank, and ask them to help you help youngsters . . . they're sure to cooperate.

- Involve your students in choosing a place to go and planning the trip.

- When you call to make arrangements, it will help if you're prepared to explain the purpose of the trip and the kinds of things the students will be looking for.

- After the field trip, use some class time to share ideas about what was seen and learned. It's always a good idea to have the students write thank-you notes to the person who arranged the visit as well as to other people who were especially helpful.
- A detailed plan of a field trip to a supermarket is described in Chapter 8. This can be used as a model for field trips to other places.

An Alternative to Guest Speakers and Field Trips. With a little help from your local telephone company, your entire class can talk to a busy bank president, a far-off senator, even a famous celebrity, or someone else who would otherwise not have the time to come to your class.

The phone company has equipment that will magnify the sound of the person's voice so the whole class can hear it; they can also provide equipment that allows the students to talk to the speaker from anywhere in the room. The bad news is, you'll have to schedule this far in advance. The good news is, the service is usually free! (See Guest Speakers and Field Trips above for additional suggestions.)

Real-World Materials Chart. The following chart lists—in alphabetical order—the real-world materials needed for the activities in this book. The chart tells you which activities use each of the materials, and gives page numbers of blackline masters for materials provided in the text. It also gives page numbers of specific suggestions regarding each of the materials. These page numbers refer to the following section, Specific Suggestions for Materials and Resources, which lists the materials in the same order as the chart and suggests ways of obtaining and using each material and related guest speakers and field trips. For easy reference, the chart is repeated on the inside front cover.

Real World Materials Chart

Materials	Blackline Master (page)	Activity or Project Number	Specific Suggestions (page)
1. Advertisements	#	1-7, 5-3, 5-5, 5-7, 8-2	262
2. Catalogs (page and order form)	41 and 42	2-8, 2-9, 3-1, 5-7, 5-8, 8-6	263
3. Checkbook Items (checks, register, and checkbook cover)	81, 85, and 88	4-2, 4-3	264
4. Credit Slips	180	7-5	265
5. Fast-Food Materials (forms and price signs)	147, 148, 149, 150	6-2	265
6. Invoices	115	5-3	266
7. Labels (bottle, box, and can labels; meat and cheese labels)	#	5-4	267
8. Magazines	#	2-9, 3-1, 8-2	267
9. Maps	#	4-7, 7-7	268
10. Newspapers	#	3-2, 4-5*, 4-8, 5-3, 5-5*, 5-7*, 7-1, 8-2	268
11. Postage Rates Charts	233	8-5	269
12. Recipes	154 and 198	6-5, 6-8H, 8-1	270
13. Register Tapes	#	5-2, 8-2	271
14. Restaurant Items (menus and order forms)	143 and 144	1-1*, 6-1, 6-7	272
15. Stamps	64	3-5*, 8-5	273
16. Supermarket Items (bottles, boxes, cans, and labels; grocery section, produce section)	# 129 130	1-7*, 2-6, 5-4, 5-5, 5-9, 8-2	274
17. Tax Tables	107	4-9D, 5-3, 6-1, 6-2	275
18. Timetables (bus, plane, and train)	175 #	7-1*, 7-4, 7-6*	276
19. TV Schedules	#	1-8	277

Easy to obtain locally

* These activities do not require the use of the listed material, but using it might enhance the real-world nature of the activity.

Specific Suggestions for Materials and Resources

This section is a treasure-trove of ideas about how to enliven your math classes with the materials and resources in the list above. We invite you to use these inexpensive, easy-to-obtain, yet interesting materials and resources in the activities and projects we've outlined in Chapters 1 through 8—and then use your imagination and invent your own real-world activities!

1. ADVERTISEMENTS

    A. *Ways to Get Materials:*

        1. Newspapers or weekly shopping ads are a great source of advertisements printed in a style that can be easily duplicated. For ways to get a classroom set of newspapers so everyone in the class can have the same advertisements, see NEWSPAPERS (10) below.

        2. If you're a camera buff, take some pictures (slides) of various advertisements, billboards, and store windows around town. You'll definitely have the attention of your class when you use them to start a math lesson. Ask questions that relate to the numbers, prices, and geometric shapes they see.

    B. *Ways to Use Materials:*

        1. Please see

Activity 1-7	Charting the Average Price	page 14
Activity 5-3	Inside an Invoice	page 114
Activity 5-5	What's the Best Buy? (VARIATION)	page 118
Activity 5-7	The Hundred-Dollar Daydream	page 122
Project 8-2	A First-Class Supermarket	page 199

        2. If you want to use ads from your local paper, use liquid correction fluid to change numbers on the ads to ones that are appropriate for your class. If you're going to duplicate more than one page of ads, run them off on different colors of paper to help keep them separate.

    C. *Guest Speakers:* Advertising executive; commercial artist; local newspaper's advertising staff; salesperson.

    D. *Field Trips:* Local newspaper; neighborhood walk to look at billboards and other forms of advertising.

2. CATALOGS

A. *Ways to Get Materials:*

1. Use the catalog page and order form on pages 41 and 42.

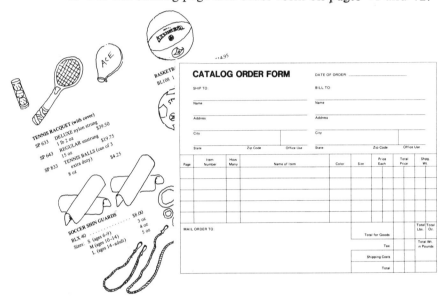

2. If you want a set of catalogs from a mail order company, find out when their current one is going to expire. Then ask the catalog store if they will save the old ones for you when the new issue comes out. If you tell them they'll be helping kids learn math . . . they won't be able to resist.

3. Ask parents to save their catalogs for you. Use the sample letter on page 257.

4. Duplicate one or more catalog pages for every student.

B. *Ways to Use Materials:*

1. Please see

C. *Guest Speakers:* Catalog store manager; order clerk or others who process mail orders; warehouse employee.

D. *Field Trips:* Local catalog store; mail order warehouse.

3. CHECKBOOK ITEMS

A. *Ways to Get Materials:*

1. Use the checks on page 81 and the check register on page 85. Duplicating the checks on colored paper makes them look more like real checks!

2. Some banks will give you sample checks; however, they usually stamp CANCELED on them, which takes away from their authentic look. (You're probably better off using ours or designing your own.)

B. *Ways to Use Materials:*

1. Please see
   Activity 4-2   Checking Up!   page 80
   Activity 4-3   A Balancing Act   page 83

2. When students need to solve money problems in their textbooks, have them write a check for each answer. You can also have them practice calculating a bank balance by giving them a check register, allotting them a beginning balance, and asking them to keep a running total after writing each check. (This activity can be self-checking if their beginning bank balance equals the sum of all the answers to the problems. When they've done all the problems they should have a bank balance of $0.)

3. Money games such as Monopoly can be played using checks and a check register instead of play money.

4. Kids really like making their own checkbooks (see page 86). All they need are several pages of checks and check registers (pages 81 and 85), a pair of scissors, and some construction paper. After they cut out the checks and check registers, they staple the check registers in the center, then fold them. They staple the checks together. For a final touch, they can design a cover to hold the checks and check register.

C. *Guest Speakers:* Bank executive; bank teller; cashier.

D. *Field trips:* Branch bank; main bank offices.

4. CREDIT SLIPS

A. *Ways to Get Materials:*

1. Use the gas credit slips on page 180.

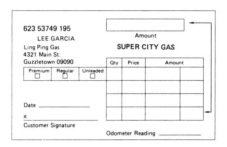

2. Your own bills are a great source for this. However, you might want to white-out your account number before you duplicate them. For some purposes, you may also want to white-out the amounts of the bills. Credit slips for school supplies, books, or other materials you've bought for the class may be especially interesting to the students.

B. *Ways to Use Materials:*

1. Please see
   Activity 7-5   Charge on Down the Road   page 177

C. *Guest Speakers:* Store clerk; financial advisor.

D. *Field Trips:* Local credit agency; neighborhood bank.

5. FAST-FOOD MATERIALS *(See also RESTAURANT ITEMS)*

A. *Ways to Get Materials:*

1. Use the forms with prices on page 147, forms without prices on page 148, and price signs on pages 149 and 150.

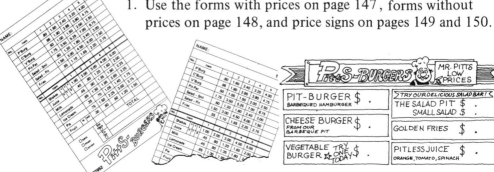

2. Ask a local fast-food restaurant for copies of the forms they use when taking orders. They will usually give you all you need. And, by the way, the best time to ask is when the prices go up. What can they possibly do with all those outdated forms?!

B. *Ways to Use Materials:*

   1. Please see
      Activity 6-2 Fast-Food Finances   page 145

   2. When the prices go up—and you have copies of both the new and the old forms—have the students compare prices.

C. *Guest Speakers:* Fast-food restaurant owner; fast-food clerk. (Many times high school students work in fast-food restaurants. You might find a speaker among the ranks of the brothers and sisters of your students.)

D. *Field Trips:* Fast-food restaurant. There are several fast-food chains whose policy is to welcome schools to visit. Some even provide a *free* hamburger, french fries, and a drink for *each* student. You'll be the most popular teacher in town!

5. **INVOICES**

   A. *Ways to Get Materials:*

      1. Use the blank invoices on page 115.

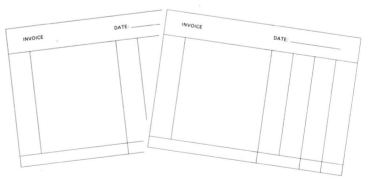

      2. You can purchase a pad of 100 blank invoices at your local variety store, discount store, or even some drugstores. You can also get some that local stores actually use—many times with their names printed on them. Duplicate the invoices if you don't have enough to go around.

3. Completed invoices are also good for math lessons. School invoices or purchase orders showing supplies ordered for your school or class will be especially interesting to your students.

B. *Ways to Use Materials:*

1. Please see
   Activity 5-3 Inside an Invoice   page 114

C. *Guest Speakers:* Clerk from a local store; store manager.

D. *Field Trips:* Department store; specialty stores.

7. LABELS

A. *Ways to Get Materials:*

1. Save labels from bottles, boxes, and packages. Look for clear, crisp labels that will be easy to duplicate.
2. Ask a friendly grocer (when the store's not too busy) to help. Tell him why you need the labels and maybe he'll have extra ones printed for you. Most supermarkets have labeling machines for their meat packages and their grocery products.

B. *Ways to Use Materials:*

1. Please see
   Activity 5-4 Meat Labels Make
                 Mean Math Lessons!   page 117
2. Paste several labels on a worksheet. White-out one piece of information (either the weight, the price per pound, or the cost of the item) before you duplicate the worksheet. Have the students find the missing numbers.
3. Have students design their own labels and make up related math problems.

C. *Guest Speakers:* Manager of the meat or dairy section in a supermarket; buyer in a department store.

D. *Field Trips:* Department store; health food store; supermarket.

8. MAGAZINES

A. *Ways to Get Materials:*

1. The first place to try is your own attic, basement, or storage room!

2. The second place to try is your friend's, neighbor's, or cousin Alex's attic, basement, or storage room.

3. For a third try—ask parents. See the sample letter on page 257.

B. *Ways to Use Materials:*

1. Please see

Activity 2-9	The Metric Marketplace Game	page 43
Activity 3-1	Geometry for the Art Gallery	page 51
Project 8-2	A First-Class Supermarket	page 199

2. Always keep a stack of magazines on hand. They make great reading material—and an even better source of pictures to cut out for number collages and other math-art projects.

C. *Guest Speakers:* Commercial artist; local magazine staff member; magazine executive; photographer.

D. *Field Trips:* Magazine offices; printing company.

9. MAPS

A. *Ways to Get Materials:*

1. A real estate office, the chamber of commerce, or the automobile association are all great sources for maps of your hometown.

2. If you want everyone to use the same map, duplicate the portion you want to use in class.

B. *Ways to Use Materials:*

1. Please see

Activity 7-7	Map-ematics Fun	page 185

2. Students use a map to practice giving and following directions.

C. *Guest Speakers:* Automobile association personnel; a colleague or friend who's a map enthusiast.

D. *Field Trips:* Automobile association; neighborhood walk to look at the lay of the land in relationship to a map.

10. NEWSPAPERS

A. *Ways to Get Materials:*

1. Day-old newspapers are easy to get. Ask the local newspaper if they'll save yesterday's paper for you.

2. Duplicate part of a newspaper so everyone in the class can have a copy of the same section.

B. *Ways to Use Materials:*

1. Please see

2. Use liquid correction fluid to change numbers to make them appropriate for your class.

C. *Guest Speakers:* Advertising executive; journalist; local newspaper staff member.

D. *Field Trips:* Local newspaper; advertising agency.

11. POSTAGE RATES CHARTS

A. *Ways to Get Materials:*

1. Use the postage rates charts on page 233.

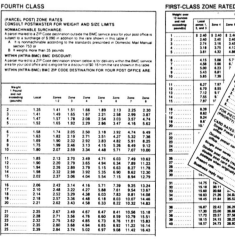

2. Ask for postage rates charts at your local post office.

3. Request rate charts from a commercial delivery service (Emery Air Freight, Federal Express, Purolator Courier, United Parcel Service, a local shipping/mail service, etc.)

4. Look in mail order catalogs or stop at a mail order store to get copies of their rate charts.

5. If you are unable to get a classroom set, you can duplicate one to make enough copies for the class. Run each type of chart on a different color paper to help keep them separate.

B. *Ways to Use Materials:*

1. Please see

Project 8-5   Classroom Post Office   page 227

2. Have your students weigh an item and then calculate how much it would cost to mail it to a given postal zone. Students can also compare the cost of mailing it first class and fourth class.

3. Have your students compare the cost of sending items by different commercial delivery services.

4. Around holiday time, have students tell about some of the gifts their family will be sending to friends or relatives. As an assignment, have some of them find the weight and destination of some packages that will be sent. The class can calculate the cost of sending each based on one or more of the following factors: speed, cost, safety, and convenience.

C. *Guest Speakers:* Driver for one of the delivery services; local postmaster; mailroom employee; messenger; shipping clerk.

D. *Field Trips:* Post office; delivery service; mail order house; messenger service; secretarial service; shipping department or mailroom of a local business, store, or manufacturer.

12. RECIPES

A. *Ways to Get Materials:*

1. Use any cookbook. (Students might find it interesting if you use a collection of recipes published by a local club or community service organization.)

2. Ask your students to bring their favorite recipe from home. (See the sample letter on page 257.)

B. *Ways to Use Materials:*

    1. Please see

C. *Guest Speakers:* Caterer; chef; short-order cook; a grandparent who loves to cook (!).

D. *Field Trips:* Bakery; catering shop; restaurant.

13. REGISTER TAPES

A. *Ways to Get Materials:*

    1. Ask your neighbors, friends, uncle-in-law, and anyone else to save them for you.

    2. Here's a way to get tapes that have only two or three addends on them. Go shopping at a time when the stores are not busy. Then when you get to the checkout stand with your 20 items, say to the cashier, "Please ring up these two items separately . . . and these two items . . . and now these two items . . . (!)" (By the way, if you tell the cashier why you need ten tapes with two items on each—guaranteed, he won't mind at all.)

B. *Ways to Use Materials:*

   1. Please see

     Activity 5-2 Cash Register Tape Arithmetic   page 112

     Project 8-2   A First-Class Supermarket   page 199

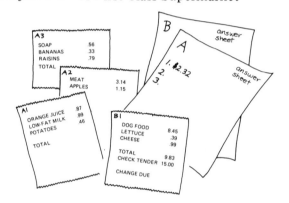

C. *Guest Speakers:* Cashier; store manager.

D. *Field Trips:* Department store; shopping center; supermarket.

14. RESTAURANT ITEMS *(See also FAST-FOOD MATERIALS)*

  A. *Ways to Get Materials:*

    1. Use the menu on page 143.

2. Take-out or travel menus from local restaurants are easy to get—especially if you say the magic words, "Help me help kids!" And don't forget to ask for a pad of the blank order forms on which the food orders are written.

3. Some restaurants have all or part of their menu printed on paper placemats. If you get one for each student's desk, it makes a lesson especially fun.

4. If you have only one copy of the menu you want, duplicate it for the class. Colored paper makes it look more realistic. You can have students paste the menu on a half piece of construction paper, then fold over the other half to make an attractive cover.

B. *Ways to Use Materials:*

1. Please see

Activity 1-1	Restaurant Bar Graph	page 5
Activity 6-1	Menu Math	page 139
Activity 6-7	The Restaurant Game	page 161

2. The next time you eat at your favorite restaurant, ask for a copy of your actual bill—and a copy of the menu. Then make copies and have the class check your bill. (Beware: Next time a waiter makes a mistake on your bill, you'll be delighted! It'll make a great math lesson.)

C. *Guest Speakers:* Cashier; cook; maitre d'; restaurant owner; waitress/waiter.

D. *Field Trips:* Local restaurant; restaurant supply house.

15. **STAMPS**

A. *Ways to Get Materials:*

1. Use the stamps on page 64.

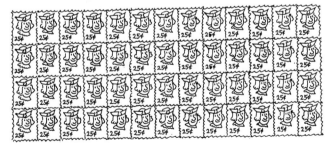

2. Inexpensive stickers that look like stamps can be purchased from school supply or stationery stores.

B. *Ways to Use Materials:*

1. Please see

Activity 3-5   Stamp-Out Game   page 61
Project 8-5   Classroom Post Office   page 227

C. *Guest Speakers:* Letter carrier; local postmaster; a favorite uncle who collects stamps (!).

D. *Field Trips:* Post office

16. SUPERMARKET ITEMS

A. *Ways to Get Materials:*

1. You can easily get empty bottles, boxes, cans, and sacks by putting out a call to parents. Use the sample letter on page 257. Ask them to cut out the *bottoms* of the packages when removing the contents. This way the empty container looks full when it sits on a shelf!

2. Newspapers are a great source of food advertisements that are printed in a style that is easy to duplicate. For ways to get a classroom set of newspapers, see NEWSPAPERS (10) above. If you want to change numbers to ones that are appropriate for your class, white-out the numbers and insert new ones before duplicating an ad for the class.

3. Use the supermarket drawings on pages 129 and 130.

4. Do you have a camera? You'll really have your students' attention if you take pictures (slides) of various shelves and sections of a local supermarket. Be sure the prices are visible in some of the slides. You can ask questions about the numbers, prices, measurements, and even the shapes they see.

B. *Ways to Use Materials:*

1. Please see

C. *Guest Speakers:* Cashier; checkout bagger; supermarket manager.

D. *Field Trips:* Supermarket.

For details of a class field trip to a supermarket, including the planning steps and follow-up, please see Chapter 8, Project 3, page 202, Behind Supermarket Scenes: A Field Trip.

17. TAX TABLES

A. *Ways to Get Materials:*

1. Use the 5 Percent Tax Table on page 107.

5 PERCENT TAX TABLE

TRANSACTION	TAX	TRANSACTION	TAX
.01 – .10	.00	.90 – 1.09	.05
.11 – .27	.01	1.10 – 1.29	.06
.28 – .47	.02	1.30 – 1.49	.07
.48 – .68	.03	1.50 – 1.69	.08
.69 – .89	.04	1.70 – 1.89	.09
1.90 – 2.09	.10	2.90 – 3.09	.15
2.10 – 2.29	.11	3.10 – 3.29	.16
2.30 – 2.49	.12	3.30 – 3.49	.17
2.50 – 2.69	.13	3.50 – 3.69	.18
2.70 – 2.89	.14	3.70 – 3.89	.19

2. Obtain a tax table from a local store, tax office, or city clerk.

B. *Ways to Use Materials:*

1. Please see

Activity 4-9D	This Taxes My Brain!	page 103
Activity 5-3	Inside an Invoice	page 114
Activity 6-1	Menu Math	page 139
Activity 6-2	Fast-Food Finances	page 145

C. *Guest Speakers:* Accountant; cashier; tax collector.

D. *Field Trips:* City or county tax collector's office; department store.

18. TIMETABLES

A. *Ways to Get Materials:*

1. Use the bus timetable on page 175.

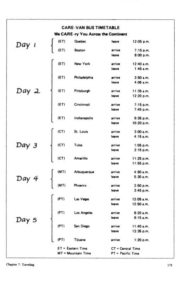

2. Timetables are readily available at bus depots, train stations, airports, and travel agencies.

3. You can also duplicate one to make enough copies for the class.

B. *Ways to Use Materials:*

1. Please see

Activity 7-1	Let's Take a Conservation Vacation	page 169
Activity 7-4	Time for a Travel Game	page 174
Activity 7-6	Bus, Train, or Plane?	page 181

C. *Guest Speakers:* Bus, train, or airline personnel; person who travels; travel agent.

D. *Field Trips:* Bus station, train depot, or airport; travel agency.

19. TV SCHEDULES

A. *Ways to Get Materials:*

1. Send a letter to parents (page 257) asking them to save this week's *TV Guide* and/or the television supplement to the Sunday paper. The best time to send the letter home is Friday, asking them to send it to school on Monday.

2. In some towns, local businesses publish an advertising brochure that also contains a TV schedule. Check your local supermarket or drug and discount centers for them. Sometimes you can get enough for everyone in your class.

3. If all else fails, you can always use your own to make copies of one or more pages.

B. *Ways to Use Materials:*

1. Please see
   Activity 1-8 Circle Graphs Make
                 Prime-Time Viewing   page 16

2. Students can use these to make up problems about time.

C. *Guest Speakers:* TV or radio station announcer, programmer, or engineer.

D. *Field Trips:* Local TV or radio station.

List of Real-World Math Games

Teaching Aids

Game Spinner (Use for Activities 1-2, 5-1, 5-10, 6-7, 7-4.)

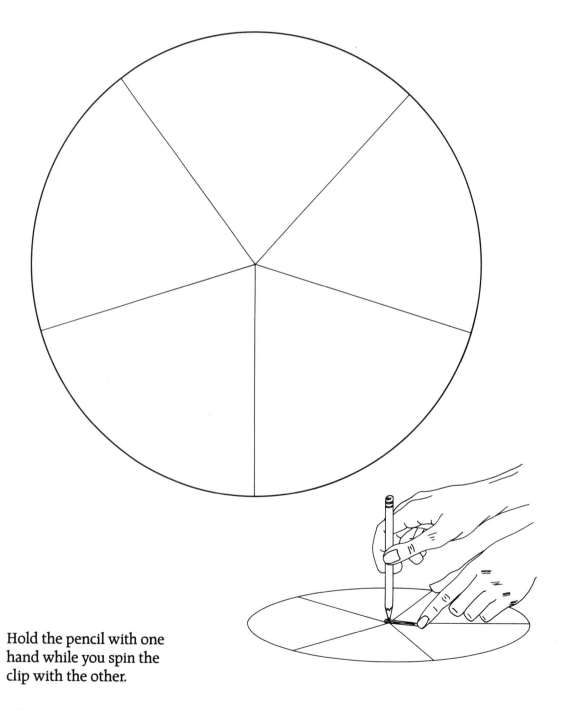

Hold the pencil with one
hand while you spin the
clip with the other.

Centimeter Rulers
(Use for Activities 2-1, 2-2, 2-4, 6-3.)

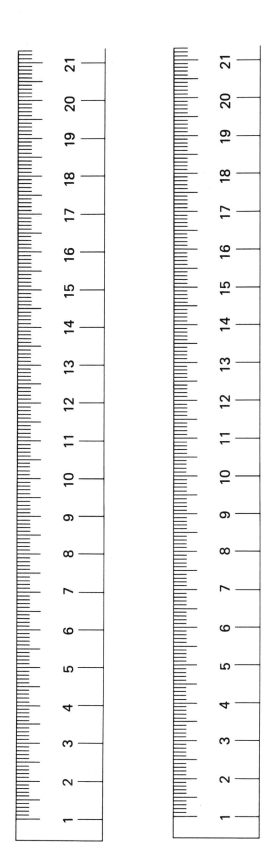
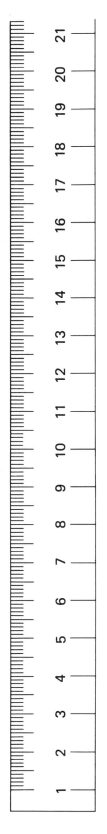

Game Board
(Use for Activities 2-9, 5-10, 6-7.)

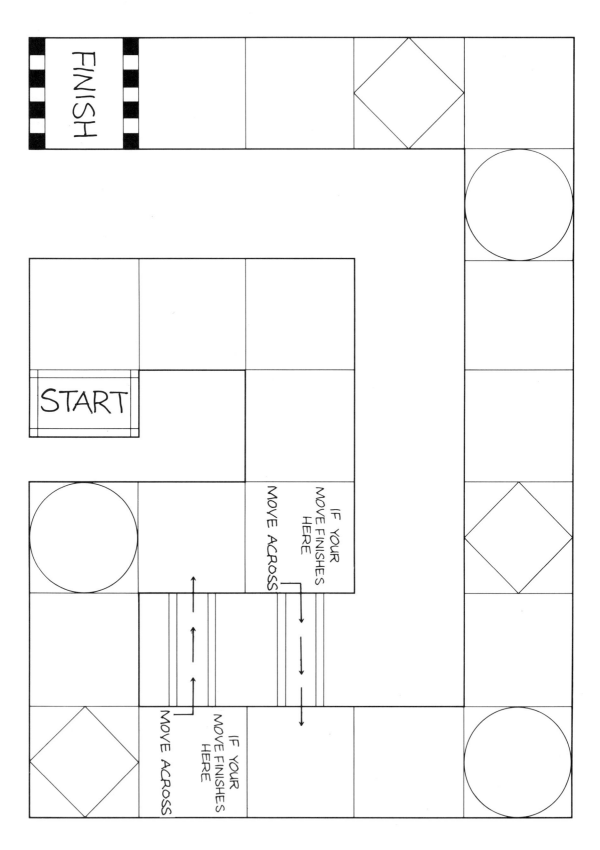

1-cm Graph Paper
(Use for Activities 1-1, 1-2, 1-9, 3-3, 3-4, 5-6, 7-3.)

5-mm Graph Paper
(Use for Activities 3-4, 6-3.)

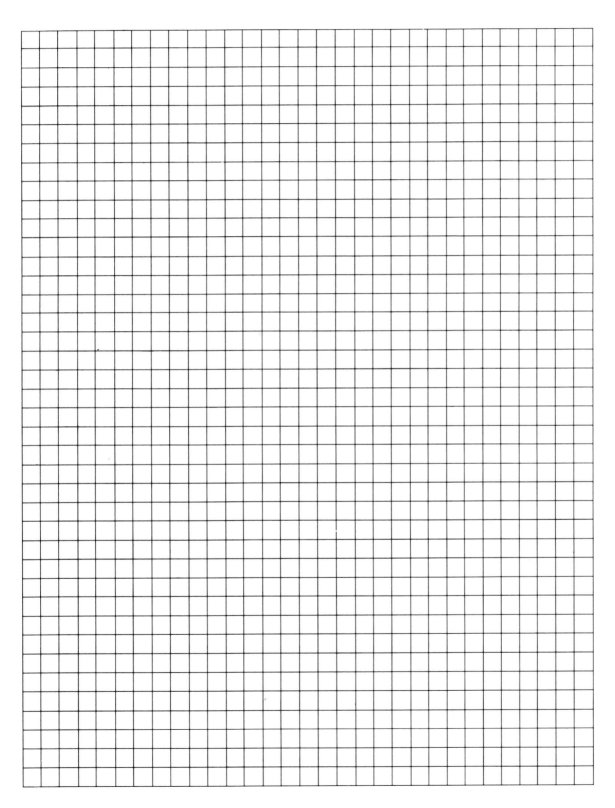

Award Certificates

REAL-WORLD MATHEMATICS
CERTIFICATE

NAME _____

DATE _____

TEACHER _____

BEANS

PEANUT BUTTER

MUSTARD

WISE CONSUMER AWARD

Student _____

Date _____

Teacher _____

FOR A JOB WELL DONE!

Happiness is Real-World Math!